T0065291

SOLOVKI'S ERSATZ

Solovki's Ersatz

(On the Evolution of Modern Human Brain)

By

Dan M. Mrejeru
2015

authorHOUSE®

AuthorHouse™
1663 Liberty Drive
Bloomington, IN 47403
www.authorhouse.com
Phone: 1 (800) 839-8640

Published by AuthorHouse 08/28/2015

ISBN: 978-1-5049-3442-8 (sc)
ISBN: 978-1-5049-3441-1 (e)

CONTENTS

PROLOGUE

Reality is not as it is, but as we interpret. Humans are caged, not freed, by the logic of their own thought that conceal everyone behind an egocentric self. Emotions and intuition are created by the magic of a golden irrational that is ratio and balance. The irrational separates number from magnitude, while this magnitude cannot be found. Logical, rational thought is the sole expression of our civilization, a starts from somewhere, follows a logical path, and concludes because all ideas have an end in themselves. How close to this conclusion are we? Is the rational world only a distorted projection of the irrational world? Does civilization make sense only for a linear mind?

When *I was a teenager I was attracted to the mysteries of the world, such as the consistency and laws of the universe, the mysteries of human evolution and all other mysteries. I was in love with Plato story on Atlantis. Later on I learned that he laid down the fundament of the logics and linearity. Then I discovered that I was a nonlinear person.*

There have been a myriad of such mysteries, and many originated from very old myths. I knew every myth was a fractal (or a grain of sand), and therefore somehow it was based on a true story (or it was self-similar to the truth). Gradually I begun to understand that most mysteries are within, or to say, they come from the way we analyze them. Here it was the reign of logic, and I had to fight against its tyrannical aspect that suppressed all deviant or lateral ideas. Then I returned to Plato just to find out how antithetical he was to nonlinearity. In contrast I found Lao Tzu, while at modern Chinese philosophers I found the concept of reversed thinking logic, the yoyo (the eddy motions) spinning system concept (in Ou Yang, Yi Lin, Wu), the cyclical time concept.

As a grownup I wanted to somehow deal with these mysteries, and I found out that the mystery itself is the nonlinear aspect of the problem. *This came in contrast with the "reality" we daily seem to know, to predict and control. I learned that people who praise authorities in all aspects, they are in fact praising ignorance. Why was this discrepancy? It occurred to me that it was there because we think this way. Could this have a scientific causality?*

For some years I tried to solve this puzzle for myself, and in the process I found out that human thinking strategy greatly changed quite recently, or in a few thousands of years. What did the change that transformed all past thinking into a mystery? When this change occur and why?

In the last year I came across information that opened a new perspective over the entire matter of interest and there was a hidden promise that might greatly contribute to my original goal. Then in January 2015 I decided to give it a new try.

It all started from the project for the Part Two of a book I wrote on creativity. Trying to identify creativity, I ran into several definitions of nonlinear versus linear, and here creativity was in lateral thinking, or nonlinear thinking. Then I realized that linear thinking developed in the human brain due to some switching of some previous nonlinear habits under genetic mutation pressure (new mutations stored inside the left hemisphere).

***

The nonlinear thinking was the strategy of hunter-gathered people built over millennia. The most important aspect of nonlinear thinking was its tendency of thinking in multiple directions, and continually searching for a starting point with the best fit. For this reason, during Paleolithic, we might have had several starting points with the same many abandons. Or even we might have had many genuine linear paths, which were short cycles, and all had a limited potential that was exhausted at some point.

Hence I thought that nonlinear thinking was based on an environmental investigation for the identification of a fact of reality from observation. Then, at some point the fact of reality that allowed to make assumptions, turned into premises. These new elements in human thinking excited the left hemisphere abilities and gradually became the logics that created a new structure of thinking. The logics generated thinking patterns, eliminated the role of sentiments, and then the linear thinking was set up in full swing. Then, the new linear strategy of thinking found its own STARTING POINT that is the starting point of what we call CIVILIZATION.

***

I was totally surprised by my conclusion, while realizing that the totality of biological systems on this planet still have a strategy of incessantly searching for the point from where to start something more complex. As it appears, on this planet only the humans came with this type of solution that generated a distinct layer of reality, a linear structure that is limited by its linearity and that is an artificiality. The linear solution is limited in time and space, and it ends when no more linear solutions are available because everything remaining is nonlinear and impenetrable.

Author and filmmaker Jason Gregory said (June 2013): "If a species built its perspective of life only within a linear conception, that very species would naturally have a propensity to fall into an artificial disposition. Could we say that the human race has fallen into this artificial disposition?" He concluded: "Culture and society, and all other apparatuses, such as government and politics, are all bull on the maintenance of the linear concept of the world. So, culture, society and everything that holds them together are artificial because nature's constituents abide by the nonlinear realm of the cosmos."

What is this artificial world? It is the place where artificial things may imitate appearances of natural things, while lacking in many respects the reality of natural. Hence an "artificial" is only an "interface" between an inner and an outer environment, while here we cannot control any of these two environments. Because the molecular world is relatively stable, the interface fulfills its purpose.

How do we build this artificiality? Human design insulates the inner environment from the outer, so an invariant relationship is created and protected to maintain the insulation against outer and inner variations. In all, we control as much as we can the invariance of the interface.

The sum of these interfaces creates a layer that is our artificial world of linear essence. In this world we try to isolate ourselves, thus no environmental variation can affect us. But this isolation kills natural adaptation and selection. For this sake our evolution ceased to occur. Humanity is in a steady state for the last several decades, where the only evolving element is technological environment by itself. It generates a bilateral relationship between people and technology, as the sole relationship influencing humanity at present.

***

We do not usually think about it, but we live in a finite world because it is linear and quantitative in fundament. At some point, in a finite world, we reach limits. Albert Einstein said: "Without changing our patent of thought, we will not be able to solve the problems we created with our current pattern of thought."

There are population limits, food limits, energy limits and cost limits. Most organizations of our worldwide societies, economical, educational, social, institutional, technological, etc., have a politico-financial bias to conserve the previous successful ways of seeing, thinking, learning and acting.

Why are the linear solutions limited? It is so because linear solutions are unique, numerical and quantitatively limited. A linear system growth has two alternatives: it can explode exponentially, or it dies out at an exponential dumping rate. A linear dynamic system is a decay of any initial present motion. The absurdities of the linear can be removed by introducing nonlinearities.

The case of space limitation is much simpler because it refers to the limits of this planet, which are obvious.

***

After many years of reading science papers from various science domains, my own research led me to the idea that the entire issue is solely connected to the production of what we call "civilization" and everything is hidden in our own brain workings.

In the meantime I found a geomagnetic excursion, named Solovki, that occurred in the epoch 7,000-4,500 years ago and which corresponds with the "creation" of, or transition to this "linear civilization" that, by all means, is a fake, or a replacement, or an "ersatz" (in German), of the real nonlinear world that surrounds us.

Adam and Eve eat the apple of linear knowledge and have been exiled from the harmony of nonlinear but natural garden. The flood brain washed humanity of nonlinear knowledge. Could we have this interpretation?

New linearity of thinking allowed people to change the environment by focusing on content, and disassemble this content into parts, and in this way it became possible for the extraction of elemental parts needed for later assembling. And this became possible by controlling the input that was meant to generate a proportional output. Since then very little changed in our thinking and behaviors.

Our civilization is rooted in the physical world, being dominated by food, energy and many other manufactured products. From the beginning (5,000 years ago) we encountered a steady growth of human population, and of the economic activity. But during this century our civilization will enter a steady-state, where the growth will cease. Why? Because we are reaching a physical limit of our energy, a limit of our efficiency (in many domains we are already 80-90% of an ideal 100% efficiency, and even at a growth of 1% per year reaching this limit is a matter of decades). Thus leveling of population (with increases of the old ages segment of population), tremendous diminishing in efficiency, significant reduction in ranges of human activities will make the economy and civilization become dominated by activities unrelated to food, energy, and all other types of manufacturing that we have today. Even it is no name for such activities.

However, after a short while the steady-state will erode the society and economy by itself and it will bring a fast decline and degradation. Here I do not mention many other natural dangers, which might strike our civilization in response to our current irresponsible linear activities.

***

For every linear action we take, we see a single result, but in fact there is a myriad of them that we term the side effects. These side effects in the natural world are as important as the only result we see, and this is the problem; nature computes differently our action than our computation, and this natural result is a new complexity that we cannot immediately observe.

We have created a linear layer of reality, that is an artificiality, and where we protect ourselves against the natural forcing of the environment. Therefore, we are no longer subject to natural selection, meaning that our own evolution ceased *because we are left with a sole relationship with the technological environment we created; this is the only forcing factor that influences our evolution, and this is a closed circuit, not a natural one, where natural evolution does not exist anymore. Could we call this an artificial evolution, too?*

***

I sectioned the narrative in search of "civilization" in 26 chapters, which all analyze the human brain workings under various environmental influences (forcing), with the hope that in the end I will have a meaningful conclusion and a relatively good answer.

This book is about why this "civilization" occurred and its significance for this planet and for us.

I am a geologist and geographer by training and during many years of studying I succeeded to develop myself into an interdisciplinary scientist. Solovki was a geophysical event; however the cosmogenic influence on human genetics was a problem because science have seen only the ionizing, damaging aspect of this radiation. According with this idea, it was only harm, but not genetic progress or change. Only recently this analysis shown possible, while favorable genetic changes.

This text is not a neuroscience book, even though I had to rely on information provided by very many neuroscience and complex system research, but also there is information provided by history, geology, geophysics, ecology, paleo-geography, anthropology, archeology, climatology, and of course the genetics. All information provides the interdisciplinary intersectionality necessary to draw a conclusion.

1. The ecosystem evolution and the environment

The entire mystery of genus homo, as becoming a bearer of civilization, is connected with **breaching the nonlinearity of the ecosystem**. In dealing with the context of the ecosystem, at a certain moment in time, **humans begun to separate and then remove the subject of their interest from the context that is nonlinear, providing for that subject a linear analysis**. The **context means relationship and interactions; removing the subject from the context, the relationship simplifies to very few elements, and the interactions disappear**. The subject become an entity (as content) that can be **analyzed** in itself; the analyzing refers to the details of its content, or its structure.

The landscape became divided into various fields, where a content of entities (subjects) existed in each field. The subject/entity turned into a quantity to which were associated some particular properties. These quantities allowed a mathematization, and a mathematical analysis that is strictly linear.

By employing this linear approach, humans disregarded the context changes, where the meaning of the subject changes, too. This process of linearization led to "ersatz" that is an artificial linear world (or a replacement world made out of myriad approximations) we created inside the nonlinear natural world, and where the ecosystem of interactions and transformations of human culture ceased to exist.

This alienation of the ecosystem was a consequence of **gradual increasing in linear thinking that was stimulated by gradual development of the language and other factors**, and where the linear process made particular subjects (the same like in the words assigned to them) to become prevalent, and these subjects become removed from their natural context (like a word removed from a sentence or phrase), being cut from their relationship and interactions with their natural ambient (a word remove from a sentence can have a different meaning in a different sentence, or context). So, **we have gradually changed the meaning of the surrounding world, and this new meaning is what we have called the "civilization".**

None of the other primates were able to produce civilization. Hence humans by their root ancestors from homo genus made a significant difference. What was that difference?

In my opinion **the ability to produce fire and to sustain it for as long as needed represented the beginning of this transformation toward linearity.** The use of fire was the first major "tool" that differentiated us of all other primates.

As the archeologists discovered, the **use of fire** occurred sometimes 1 million years ago. But **the use of fire as a full technology became available only some 500,000-400,000 years ago or less.** However it did not turn into a powerful challenge, until other subsequent elements were added to hominin brain in the form of multiple mutations. They were new skills and they needed new neuronal structures to process them.

The most important role of fire consisted in the **ability to cook food.** Thus this cooking allowed the body to save food processing energy, and this saved energy became available for the brain that typically uses 20% of the entire body energy for its own functioning. The result of this process was called **"encephalization" because it increased the size of the brain.**

For this era of the beginning of encephalization (since 500,000-400,000 years ago) we have the Homo erectus ancestor (who migrated out of Africa about 2 million years ago and inhabited in Asia 1.8-1.7 million years ago the space from Georgia to China and Indonesia), and who around 140,000 ago reached a brain size of 1,100 ml. or cc.

Many scientists consider that brain size is not necessary the element that has brought intelligence. And here the case of many "very smart" birds and insects is usually cited.

However the oldest stone tools ever found were from Africa and dated 2.5 million years ago, but a crucial turn in tool making occurred only since

500,000 years ago and coincided with the spread of the fire technology (cooking as the most significant).

There is known that several primates use to make tools to improve their efficiency for feeding. They use wood, stone and other materials available in their environment. Hence culture and tool making is present in the same sections of the primate brain in charge with these abilities that co-evolve with the section dedicated to language.

It has been speculated that the environment powerfully influenced material culture because finding tool materials and resources asks for constructing and using the tools. The speculation insists that **here is an ecological opportunity, rather than necessity**. This ecological opportunity would be the main driver of primates. Hominins have shown in this regard behavioral flexibility and plasticity, which have influenced their material culture.

However, none of these advents produced "civilization" as we know it. More than that, many scientists denied the argument of encephalization, and of tool making, showing that it was nothing radical enough to produce "civilization". Hence all gradualities we found had not produced "civilization", had not been very relevant, but they only push humanity to the point, where it needed some tools to produce this civilization. And this phenomenon occurred only very recently, or starting with 7,000-5,000 years ago.

So **the tools were needed by the nonlinear thinkers, not by the linear thinkers who did not exist, yet**. In the rest of the text we will see how this transition from nonlinear to linear occurred, and what made the linear to succeed, or to become prevalent.

***

Let's talk a little bit about the ecosystem and the environment, which allowed this journey toward "civilization" and drove our species into the current "modernism".

The impact of geography and environment on morphological variation in animals is well recognized in biology and zoology. Body proportions, facial shape are linked to mean temperature of the environment; temperature help to determine the body size, form and proportions in hominin taxa.

There are ecogeographic variations, like rainfall and moisture, with impact on morphology.

For example, starting with H. erectus, cold-adapted hominins were exposed to extreme (cold) temperature only for a part of their existence; thus their morphology only in part reflect cold-climate adaptation. This case is well defined in cold-adaptation of much more recent haplotypes, like R1 and O3 (Mongoloids).

The ecosystem is basically a network of nodes, where the members (agents) interact. This interaction generates complimentary relationships and very little competition. The nodes are pre-programed to design "spatial regulations".

The evolution of the system generates new nodes, which are new attractors. The same influence is produced by climate changes.

The ecosystem functioning results from interactions among and within different levels of biota. Also it occurs an interaction between different types of ecosystems. The plant species moved geographically in response to environmental challenges. Native sites are not fixed geographic areas. Within diffusion-dispersal-pathway of special importance is the formation of corridors, which allow a **substantial exchange of flora and fauna. Corridors open up a large area that can be colonized rapidly,** although the colonization happen through the rather narrow connection provided by the corridor. Long-distance dispersal allows areas to be colonized that are far beyond typical ecological time scales. Here the population shortly show strong founder effect and quickly diverge from the populations in the original range. However this effect of long distance dispersal can appear on quite short range phenomena, too.

During climate changes the functioning is decreased as the number of species decreases and vice-versa for periods of stability. Disturbed environments can have long term or short term effects on plant community composition depending on the degree and duration of the disturbance event. Natural disturbing events are: drought, fire, disease, insect, glaciation, heat stress, freeze stress, volcanic smoke, soil deposition, soil salt level, animal pressure, etc. These phenomena produced **pre-anthropological intercontinental plant migrations**.

Here we have to mention that on Earth manifest many cyclic phenomena, which all affects the landscape, environments and ecosystems.

These cycles mainly influence the climate, but many other types of influences occur, too.

Many organisms have habitat preferences, such as particular types of vegetation, preferred temperatures and precipitation. When changes appear to these aforementioned elements, the animals can move and track their favored habitat or adapt by genetic changes (mutations); otherwise they become extinct.

In the case of homo, it is assumed that many physical changes (over very long periods of time) have appeared in connection to a particular adaptation to new environmental conditions (like upright walking, climbing ability, swimming, specific tool making, etc.). **Environmental change in one direction can lead to specialization for those particular conditions**. The variation in behavior of the individual has helped survive variable changes, without relying on specialization. This has caused the aforementioned **plasticity of behavior**.

Hence **many key adaptations have appeared in response to** selection pressure imposed by the environmental variability and instability. Homo advantage has been his capacity to adapt his behavior to changing habitats and conditions, instead of specializing for a single type of environment. This capacity has helped him to coop with environmental diversity and ultimately to migrate long distance. Several scientists consider that homo had **accumulated a** *"pool of adaptations, which were not expressed immediately, but only under the pressure of selection, and when a particular adaptation was selected out of the pool"*.

Some of these pressures were cyclic in nature. For example the most important cycles are connected to 11-years solar cyclicity and 22-years solar cyclicity. We have also Gleisberg solar cycle of 60+/-10 years. At present we know that global wheat production follows 11-years solar cycle. Hence the same cyclicity have influenced in the past the plant efficiency of every environment, causing inside disturbances and movement of animals.

Periods of higher solar activity and lower cosmic ray flux are associated to warmer climate and vice-versa. Lower geomagnetic intensity is associated to cooling of climate. Large volcanic events also have cooling effects. Global volcanism appears to have a quasi-periodic behavior in the range of 15+/-5 years. Global seismicity has 30+/-10 years periodicity.

5

An analysis of volcanism shows that the most powerful volcanic events are clustering around geomagnetic jerks events. Jerks are lasting 1-3 years, but some occur in groups of quick succession. Dipole part of geomagnetic field generates global jerks, and the non-dipole part generates local jerks.

However we estimate that many more cycles exist in nature, and all of them have certain influence on environments and ecosystems because all environments are immersed in and attuned to all these cycles.

In general **during the Ice Ages**, the activity of elements is higher than in the inter-glacial eras because the **polarization is amplified by discrepant factors, like glaciated versus none glaciated environments**.

Thus the migration of genus homo into Eurasia had provided very many challenges due to the fact that the ecosystems there were distinct and implied new strategies of surviving. However the main differentiation was in the higher elevation of the local grounds and the mountainous structure of the landscape, where the exposure to radiation (cosmogenic or solar) was also higher than in the plains and at the sea level.

For example several settlements found in the Peruvian Andes at 4,300-4,500 m altitude were dated 12,800-12,400 years old. It was assumed that H. sapiens entered Americas from Siberia, crossing Beringia, some 15,000-14,000 years ago. Therefore, the adaptation to high altitude of Peruvian Andes occurred quite fast (probably during a time interval of less than 1,000 years).

All these challenges forced into self-organization that produced adaptation. During migration out of Africa, the rate of mutation was higher for H. sapiens than within previous African habitat, and the forcing factor was the need to adapt to new types of environment.

***

Let's see other features of the ecosystem mechanism, which have influenced the H. sapiens migrants.

Natural ecosystems create agglomerations, clusters and patterns. The diffusion in the ecosystem tends to assure uniform population density. Sometimes diffusion-induced instability generates **spatial patterns** (of higher or lesser density), **which become new nodes of the network**. The immersion of homo into new ecosystems can alter the rate of diffusion. When everything in the ecosystem is developed at reasonably similar density, no new centers will emerge beyond the existing ones, unless this stability is not disturbed by the migrating agents.

The boundary of the ecosystem will disperse to provide optimum interacting distance between nodes and agents, when new agents enter the system or other factors intervene to enlarge the membership. The network is regulated by self-organization that regulates the ecosystem needs by adaptive measures.

However genus homo was able to exit an original ecosystem and to diffuse into neighboring environments. This diffusion was caused by climate changes and other local challenges.

Within the ecosystem two forces act: centripetal and centrifugal. Centripetal is self-organization itself. Centrifugal is not a force but a concept that represents an effect; hence it is an imagined force. It is a reaction: the reaction to centripetal is centrifugal.

Originally, genus homo (hominins) had an attractor of his own ecosystem. The collapse of this ecosystem, mostly due to climate changes, made homo to cross (disperse out) the ecosystem boundary into other ecosystems. This dispersion was caused by the boundary collapse and followed an attraction present in other ecosystems.

The **dispersal was possible only by adaptation, and adaptation implied new neuronal circuits. Since the process of diffusion in neighboring ecosystems repeated itself with a certain frequency, more new neuronal circuits were created**. In each new ecosystem, genus homo had to deal with local attraction, and this attraction produced adaptation to new environmental conditions. New environment had distinct identity, so homo adapted to this new identity and other ecosystem patterns.

There is to consider that the adaptation to this differentiation between ecosystems was not an instant process. On the contrary**, it took some time to ignite the self-organization process, which built new neuronal circuits and interactions**. However a determinant factor was the accumulation of mutations within a pool, like within the brain.

For example when homo reached today land of Afghanistan, excepting easy access from today Iran and Central Asia, he found himself in the situation to cross Hindu Kush range through five main passes, and only Khyber Pass had an elevation of 1,100 m because the rest of other four passes had elevation exceeding 2,700-3,000 m, and two of them, Kushan

(4,370 m) and Salang (3,878 m), will cause hypoxia for any trespasser. An even worst situation was for crossing from today Afghanistan into today Xinjiang (China), or from today India and Pakistan into today Tibet, Yunnan and Sichuan because of Himalayas passes with elevations over 4,000 meters.

However, we learn about a significant uplift in the last 2.5 million years in the Hindukush-Himalayas Mountain Ranges and on the Tibetan Plateau. For this region geological information is quite contradictory on rate of uplift for this large region of South-Central Asia various scientists have found rates from 0.3 mm to 5.8 mm for the Tibetan Plateau, but twice higher for Himalayas. **In the Tibetan Plateau the uplifting in the last 2.5 million years was 2500 meters**, but we do not know how much was in the last 50,000 years, or since H. sapiens has reached this region. Based on lowest uplift rate the plateau was risen some 60 meters, and double this figure for Himalayas. Based on the highest rate, the **uplift was 500-600 meters in the last 50,000 years, and even twice as high in the Himalayas**.

These figures can be quite important because they lower the elevation gap to be crossed by H. sapiens in order to reach north-Central Asia. For example an elevation of 3,500-4,000 m causes hypoxia, but lowering this elevation with 500-600 m the crossing becomes possible without hypoxia complication. In the meantime, **a H. sapiens or hominid living on the actively uplifting plateau, in the course of time, will naturally adapt to this situation of increased elevation**. Probably this situation was experienced by our cousins, the Denisovans, who came long before the H. sapiens in this region from Altai. However, the Altai Mountains also shown an uplifting process.

Crossing these high passes implied stage accommodation to lesser oxygen of the high altitude atmosphere. Maybe homo needed several generations of continual adaptation to finally succeed in crossing the mountains. The same mountainous regions had temperature variations of 20 centigrade during the day, with night temperature running below -20 centigrade, and at higher altitude the temperature drops to -40 centigrade. One should imagine what challenge will be for these homo from the tropics and equator to make a living in such distinct climate conditions. A similar sharp adaptation was required for those homo settled in Siberia.

We can say that the **migration was not a continual process, but it was more like a wobbling (from side to side, while back and forth) in the ecosystem space and hooping from one area to another**.

The migrants had difficulty with "starting", meaning they were uncertain when and where to move on. **These characteristics were typical for the "space thinking" present in their mind processing**.

Probably the residence into each area was at least for several generations (assuming that one generation is 20 years, the residence in one place was for 100-200 years, more or less). Now one place means an area defined by the boundary of certain ecosystem. Hence such an area can be larger or smaller (it may vary from tens of kilometers to many hundreds of kilometers), and it is defined by the configuration of the landscape. It usually is a valley with affluent rivers that is surrounded by hills or mountains. At the river confluence would be the ecosystem highest density. On the interfluvium, in the mountains, would be the lowest, and here would be the passes, or the gates toward other ecosystems. In most cases the interfluvium is the boundary of the ecosystem.

Centripetal attracts, or it is an attractor in itself. It causes centralization, increasing the density toward the center, increases the density of interaction between agents, and it is responsible for certain transformations. Centripetal creates proximity. For this sake homo needed to reach that center, where the resources were maximal.

Centrifugal represents an attraction toward another but exterior center that creates diffusion. As more diffusion happens, the dispersed area begins to re-densify, acquiring new identity.

Thus centrifugal is the opposite of all aforementioned characteristics of centripetal. In general decentralization, caused by centrifugal, is seen as a negative effect. Centrifugal can change the shape of the boundary created by centripetal.

Centripetal causes the identity of the center. Challenging and changing the identity of the center results from the influence exercised by the outside attractors, which compete with the center; the ensuing effect is seen as centrifugal, causing a dispersal of this center identity in the surrounding territory.

Following the aforementioned rules of the mechanism, genus **homo was pendulating inside each ecosystem, searching for local identities and patterns, until he became diffused toward new attraction.**

According to this scenario, **the path of migration was never a straight line, except for short paths**.

We can say that **a particular magnetic sense guided homo along these paths**. It was recently proved by scientific research (2013) that this **magnetic orientation exists in humans and it is connected with the works of the pineal gland that has a structure similar to human eye.**

The above aforementioned research might allow the speculation **that humans, guided by magneto-sensibility, followed the grid of the geomagnetic field** that is usually aligned north to south. But during the geomagnetic excursion the axis of the magnetic field flips toward east in the northern hemisphere, and toward west in the southern hemisphere. Or, it turns to be 45-60 degrees, or even more, deviated from the north-south axis. Hence, during such an event, the humans from the northern hemisphere will have the tendency to move eastward. Probably this happen during Gaotai event (80,000-70,000 years ago), causing out of Africa migration into Eurasia. Later, the returning of the axis to normal north-south position favored a movement on south-north direction.

Now, during the Laschamp short-lived geomagnetic excursion (41,000 years ago), the situation repeated with movements toward east (East Asia and Australia), followed by movements toward west (Europe) and north (to Siberia), being caused by the alleged reversal. 10,000 years later, the Mono Lake event caused similar outcomes, but with moves toward east. In short, all geomagnetic excursions produced significant human, and eventual animal-plant movement (migration).

Two forces regulate the faith of the ecosystem together with self-organization. Since the beginning of civilization, as settlements, the same ecological type of mechanism was copied in the design and rules by human society and its infrastructures. Even when the design and rules are similar to those of the ecosystems, our civilization did not created anything as an ecosystem.

Our civilization follows the behavior such as the species exhaust all environmental resources, collapsing the system. This severely questions our survivability and our approach toward civilization.

Let's return to ecosystem itself and to the cause that made genus homo to migrate all over the planet. As we have explained the miracle that ignited genus homo's voyage toward civilization was the advent of fire, and ultimately its use for cooking. A large part of the energy consumed by the body in processing the food was saved by cooking; this supplemental energy allowed the brains to use it in order to extend by adding new neuronal circuits, which were needed by new acquired abilities.

The **encephalization process endowed genus homo with unusual qualities compared with other members of the ecosystem because he was able to accumulate adaptations, abilities, experience and ultimately cognition.**

In the end**, the sum total of these qualities (abilities) were enough comprehensive to allow homo to flexibly adapt to any other ecosystem of this planet and this was his ticket for intercontinental migration.**

In sum the adaptations changed H. sapiens in significant way, and the advent of language helped the accumulation of knowledge. It was something else that we call **linearity** that may be considered as the single cause of producing "civilization". In the meantime the term "civilization" is how we define our own but recent stage of evolution; but it is only a human interpretation given to the current stage of "cognitive linearity". In the opposition to this linear approach everything around us, including ourselves, is nonlinear and complex. Thus **we have created an artificial linear construct that competes by opposing every natural ecosystem**.

The development of language encouraged humans to treat the world around them like the words from a sentence. However, when one changes the context, meaning the word is used in a different sentence, the meaning of the word changes, too. Analyzing a sentence, one finds a pattern and a meaning, which both directly depends on the **"content"** of the sentence **that is limited to particular conjectur**e.

The **analysis is a quantification of the content and** cannot bring the bigger picture because it disregards the dynamics of change, the relationships and the interactions. **The analysis unveils only a "frozen reality", or a sequence only from the ever-changing dynamics.**

The ecosystem is the ever-changing dynamics, it is a world of connectivity, relationship and interactions. **Our "artificial construction" deals with a "frozen space/reality" that projects into the future.**

One can say: change the environment and you change the intelligence. Changing the environment is changing the context, and the **intelligence has a different meaning in a different context**.

Many studies indicate that genes, determining the intelligence, exist and can be passed to offspring. Much of the intelligence we currently use is partially due to environmental and ecosystemic factors, and results from evolution caused by the outside forcing of the ambient. Much of this intelligence had a nonlinear character until recently (5,000 years, or less).

Our ancestors were forced, by changes in the environment, to adapt. Both changes and adaptations were not rational because both were driven by nonlinear dynamics. Hence, we were created to be not rational. Rational is about our view on reality, but rational changes due to context changes.

Belief is rational, but has roots in an approximation of irrational realities; this often allows the individual belief to deviate from the rational norm. Everyone holds a quantity of irrationality (every stable system is made of many instabilities). Thus irrationality is as good as every rationality, or every rationality needs some irrationality to make it workable. For example, in economics it was found that irrationality increases as monetary incentive augments.

Rationality is limited by the available information that depends on the nature of the context. For example, every distance requires an infinite amount of information to be known. Hence, what we see as space is not real. The reasoning stands for time intervals, which are not real, too. Some say that irrationals are not exactly numbers. Then, what are they? Some say, they are immaterial, or virtual points. This leads to the conclusion that the universe is immaterial. Then, what the rational numbers are? Some say that they are only a particular projection of the immaterial. How many such projections can exist? Is the universe a multiverse?

For us there is that the **minds exploit (by approximating) the apparent structural regularity of an irregular environment. It is known that the Paleolithic people used, within tool making and other constructive approaches, proportions like 2/1 (0.50), 3/1 (0.33) and 3/2 (0.66), which later were proved to be approximations of irrational numbers like pi and phi.** The prehistoric people could not make a decision because of limited computational resources, but even then, they were able to identify, by large approximation, the ever repeating natural patterns of the environment. What is the ecosystem connection to all this? It is about relationship, that being material or/and immaterial, still defines this aspect.

As a result of our confrontation with the ecosystems we have created our own artificial world, where we try to protect ourselves against any exterior forcing. By this approach **we eliminate the role of "evolution", meaning our intelligence evolution ceased inside this "artificial linearity".**

Here **we are the "captives" of our own "artificial system", or captive of our own technology.** The future development, directly and sole, depends on the technology we produce.

A study on different primates and mammalians shown that for all species the individuals in **"natural captivity" develop smaller brains and show lesser intelligence.** The study refers to a type of "natural captivity" created by the environmental conditions. **The linear thinking is the cage of our captivity.**

I cannot end this chapter without saying that a human is an ecosystem in itself. The wellbeing of this ecosystem depends on human emotions.

Charles Q. Choi, a Live Science contributor, on August 3, 2006 wrote the article: Study Reveals the Logic behind Our Rational Brains. He said: *"philosophers have conceived of a place of emotions in the topography of the mind particularly in the relation to body states. A number of theories of emotions stress their function in the conduct of life. It is a particular ambivalent relation between emotions and morality. Emotion theorists have recently turned to collective or shared emotions, as a shared form of shared intentionality. They play an indispensable role in determining the quality of life because they involve more pervasive bodily manifestations than other conscious states. They contribute crucially to defining our ends and priorities. They protect us from an excessive slavish devotion to narrow conceptions of rationality. They have a central place in moral education and moral life."*

There is no doubt that these emotions have a critical role within our very narrow, but personal ecosystem. The visual, or space thinking, as a nonlinear thinking system, was including the sentiments, emotions, and feelings in the process of thinking. **With the advent of verbal thinking this part is mostly eliminated as being irrational.** However, this irrationality originates into a proportionality of certain magnitude that is significantly defining for the brain processing and our own wellbeing.

Among the example of emotion can be selected the epistemic category (interest, curiosity, conviction, and doubt), the esthetic category, and the positive-negative category (guilt, shame, envy, disgust, and sentimentality, and on the other hand their polar opposites). For another example, the emotions in music show a relation with the rhythm. In Mathematics, the association with rational numbers show relationship with harmonics, while the association with irrational numbers implies the sharing of space.

Mathematician Yi jing explains: *"Our being is grounded in symmetry, and also in the dynamics of instincts and habits. Symmetry is a natural response to energy conservation. Emotions are ratios, or balance. Feelings cover a communication of intent, when dealing with context that helps to hide as well as assert identity. The aggregation of 'essentials' form sameness in response to context. Consciousness and language are agents of mediation. The Mathematics derivate as a language. Uncertainty is a property of mediation, while certainty is in the symmetry realm."*

2. How do we come to our modern brain?

Recent research on the white matter in the brain provide some explanation for the nonlinear abilities found in many individuals, especially those with autism. Casanova and colleagues (2006, 2007 and 2009) found that such "individuals with special **brains have more circuits (mini-columns) per square centimeter than other individuals; disadvantage is that these brain constructions have fewer long-distance connections between distant brain regions**".

However these special individuals are term geniuses, but they have low type of social behavior; on the opposite they have an extremely high ability for visual details, which are processed locally by their **higher density of neurons**. In other words **they have access to a richer "context" and richer display of interactions**. This facilitate the "context-understanding", and not only this but also they are higher innovative, or creative because they have the chance to see many more "patterns" from reality and their transformation involved in environmental processing.

Some other scientists suggest that the same "visual ability" neuronal zone have been exploited by the "verbal ability", and this is the place where the overlapping occurred, **producing the suppression of the visual abilities**. This mechanism was found in recent research on the IQ testing that discovered the high IQ scores connected to verbal thinking, while the same high IQ scoring individuals were the worst in the analysis of details.

As another example, Marvin Bartel said: *"In March, 2005, Sir Ken Robinson, chair of UK Government's report on creativity, education and economy described a research that showed that **young people lost their ability to think in divergent or nonlinear ways**. Of 1,600 children aged*

3 to 5, who were tested, 98% showed they could think in divergent ways. By the time they were aged 8 to 10, 32% could think divergently. When the same test was applied to 13 to 15 year-old, only 10% could think in this way. And when the test was used with 200,000 individuals 25-year-olds, only 2% could think divergently".

This study indicates that **human brains at an early age still preserve a natural nonlinear capability (98%) that is suppressed by education and social rules, and at the age of 25 an adult preserves no more that 2% of the natural endowment.**

This is the case of **a recent change in our brain functioning that has a history of only 200-400 generations** (covering 4,000-8,000 years interval). The **nonlinear thinking was created during a time span of at least 100,000 generations**, and it is still overwhelmingly present in our brains (98%) until we enter the education process (at the age of 3-5). The first batch of "linearity" is provided by our own parents.

According to neuroscientist Dr. Shanida Nataraja, during meditation the right brain is mainly used, and this right brain does not have the ability to categorize and analyze the meditation experience; it intuitively "feels it". This shows how our antecessors got into meditation by emptying the verbal mind, by plant ingestion and long practice of this attention concentration, and what results were obtained by this meditation performed by local shamans and priests. This research found the **attention as the main prerogative of the right brain, where all details of the context are summarized**. The meditation uses a special concentration, where these details are made to gradually disappear, and hence it totally acts and deal with the right brain. During the same process the intellectualized left brain is completely silenced as the first stage of the concentration technique.

Dr. S. Nataraja research shown that the right brain is too much emotional sensitive, and the **Neolithic and ancient people were very emotional**. However the research shown that the right brain is the most "plastic" part, meaning that all emotions were controlled, being turned away by rapid adaptation. This explains, why the pre-historic people were fast argumental people, but very rarely the argumentation turned into an opened conflict. As it appears the "conflict" itself is mostly driven by logics, and it shows an interest-driven reason for conflict. The archeologists found evidence for several local conflicts that produced mass murder within several sites in Central Europe, which were dated 7,000 to 5,000 years ago. Nothing of this nature was recorded before this time. Here we had an evidence on the use of logics.

In other words, **during our recent evolution we lost a part of our brain plasticity that was replaced by the employment of logics. The logics also replaced the process of natural evolution of our brains**.

Another study made on adolescents showed that **more naturally (open) driven evolution through education produces higher IQs** than the conservative type or controlled type of education: liberal and atheist adolescents score 106-103 points at the IQ test compared to conservative and religious adolescents who score only 95 and 97 points. The greatest difference is between liberals (106 points) and conservatives (95 points) that is the difference between open-minded and narrow-minded. Or one can say liberal education serves better the contextual processing of human brain.

Our brain evolution is responsible for the civilization we created in the last 7,000-5,000 year or less. As the research indicates, **during this interval of time, our culture increased exponentially**. The aforementioned studies found that "**creativity**" (increases in individual "creativity") **played a major role at the origin of human culture and for its accumulation through history**. Now the aforementioned studies showed that "**divergent thinking is the motor of creativity**" and it originates in the right brain. In other words, we are "here" because of our past "creativity", and our future still depends on the same "creativity".

"**Creativity**" **led to the discovery of locally adaptive information**. The model of social learning explains that when individuals imitate, the imitators will acquire the locally adaptive behavior with the same probability as those individuals who learn by their own experience and at a cost. So the chain of imitating is rooted in someone who extracted information from the environment. **Collective learning is cultural learning, but learning from others, not directly from environmental cues**. Imitation has higher fitness, and the propensity to spread. **Cultural learning can produce behaviors that are different from those provided by environmental cues**. The ability to learn culturally have raised the average fitness of the population by allowing improvements to accumulate from one generation to the next.

During the history, large changes have improved things only when they were inside like a geometry of a small cone that included the distant optimum. The same procedure applies to today discoveries. Many special discoveries extracted from the environment by divergent-thinking-individuals were preserved in the form of some secret codes, and hence no one was able to imitate them unless he/she knew the code. When the only individual, who knew the code, died without passing by the information, the said improvement (discovery) disappeared, and the offspring must begin again the same process of discovery as his predecessors.

The aforementioned aspect intermittently appeared in the epoch 24,000-5,000 years ago, or for 20,000 years, before the linearity prevalence, and it was a manifestation of nonlinear thinking and ideas. This epoch had the stone, wood, bone and other materials tools to concretize some of the major ideas of this epoch. Some scientists speculate that the maximal achievements of this epoch must be connected to regional occurrence of artificial stone discovery and production that appeared and disappeared repeatedly. It means that the artificial stone was discovered and rediscovered several times because of continual loosing of the receipt, or formula. Probably in the process of rediscovery, the formula itself changed several times.

***

During the long passage of history this linear step-by-step improvement (but caused by divergent thinking) gradually diversified, giving rise today to another exponentially growing linear diversification.

According to various estimates, the world population was 2.5 million people 10,000 BC, and it doubled to 5 million in 8,000 BC. At 6,000 BC world population doubled again to 11 million people (at the beginning of the agriculture). Again and again the population doubled to 28 million in 4,000 BC, to 72 million in 2,000 BC and 150 mil.-170 mil. in the year 1 AD. In the year 1,000 AD the world population doubled again (for the first time in half of the time, or in 1,000 years) to almost 300 million. As one can see the process showed every 2,000 years a constant doubling occurred for the last 12,000 years that include the Mesolithic, Neolithic, and Antiquity.

This statistical estimate do not show any spectacular or any uncommon increasing in the world population due to the development of agriculture. However another doubling (next) occurred in only 1,000 years in 1000 AD, and in 500 years, or around 1,500 AD when it reached 500 million people. Now, or at present, or 500 years later, the population increase is more than 10 times. Based on the above information, **the most uncommon era in human development had occurred in the last 500 years**.

***

The imitating is the job of mirror neurons and it is not unique to humans.

Now, one will wonder how do we came to this stage of evolution in our thinking system. This chapter is intended to answer this question.

But when and how our brain transformed in the manner to produce a "civilization"?

For the scope of this paper we intend to find only the breakthrough element that brought us to "civilization", the approximate epoch of this occurrence and the events, which shaped our current status.

As much as **the brain** concerns, the process of "**encephalization**" produced a larger brain with many new interneuronal connections. During evolution the cortical part of the brain increases much more than the subcortical one. Mammalian brains becomes folded inward as they increase in size.

The volume of gray cortical matter is 50% of the whole brain matter, while the size of cerebral cortex is 80% of the entire human and hominin brains. The neocortex is the largest part of the whole cerebral cortex. Here the cerebellum is only 10-15% of brain's volume but 75% of its mass. It has been observed a coordinated evolution between neocortex and cerebellum because the enlarging of cortex and/or neocortex would produce the enlargement of cerebellum.

The connection of fiber bundles travels through the white matter, and this accounts for the cortical surface. As aforementioned this surface folds inward, and this is called convolution. Because of this feature, the brain can be more compact and faster with increasing size.

The volume of cortical gray matter is a linear function of the brain volume, so it is the same proportionality that is preserved unchanged for all primates. The white matter mass (that is 35% in human brain) increases disproportionally with brain size. This means that the increases in brain size would produce more increases in white matter: for example, a 3,000 cc or ml brain would have 50% white matter (compared to 35% at our current size).

Evolutionary process of neocorticalization in primates **is due to the progressive expansion of the axonal mass** rather than increases in cortical neurons. This shows the **importance of high neural connectivity in evolution of the brain size**. Human cerebral cortex is dedicated to conceptual and not to perceptual and motor processing.

The sensory modalities have around one dozen representations for visual, and half a dozen each for auditory input and somatosensory sensations. These modalities are flexible and continually modified to provide the needed plasticity for adaptive behavior and learning.

The brain evolution is defined by the cooperative association and hierarchical organization of neural circuits. Fractal folding reduces the interconnective axonal distance and produces higher order associations. In general, evolution increased the number of circuits.

Neurons are natural migrants; most, if not all, of the neurons in the mammalian nervous system migrate from their places of birth to their location of function. Neurons usually originate in the ventricular zone, where their precursor cells proliferate. They can migrate radially to other layers in the brain, or tangentially that is parallel with the surface of the brain.

A new study (Letinic and Rakic, 2001), by comparing neuronal migration in humans with that occurring in other animals, demonstrates that **human brain possess migratory pathways that do not exist in other mammals, or other primates**. As a result, human thalamic nuclei connected to the frontal cortex are larger than those in other primates. Rakic and colleagues shown the co-evolution of frontal cortex and the thalamic nuclei, and the migration process contributed to more neurons in human thalamus.

***

Functional limit of intelligence consists in its ability to process information. A better processing is achieved by reducing the length and number of axons, and this reduces the axonal mass. This is achieved by reducing the number of fibers, while creating modular circuits. The length is reduced by fractal folding, and this reduces the white matter volume and its energy consumption. It has been calculated that the **entire human brain uses a power of 15 watts**.

The scientists have calculated the maximal admitted size of the brain expansion that must stay below 2,800 cc or ml. Over this size the capability of neural integration declines. It is considered that **human brain lies about 20-30% below the optimal capacity**, while the volume that can be achieved can be double of the actual size.

The analysis of the increased brain size has indicated some associated events like precociality, increased birth space, single births, prolonged periods of postnatal development; the delays in maturation and reproductive rates increases the "generation time", but also increase the life span (a raport of 3:2, where the first digit is the brain size increases, and the second digit represent increases in life span; so a doubling of our brain size will extend our life with 50%).

Size of 3,500 cc or ml produces the maximum processing capability. Any further enhancement will require structural changes, like improvement of neural organization and signaling process, new neurons and improved thermodynamics.

***

Frontal lobe is at the forefront of brain evolution because its genes are involved in plasticity (adaptation), but also it is devoted to behavioral and social interaction.

In total **our brain has 214 genes involved in our evolution**.

Let's take the case of Neanderthals. Their brains were on average 1,600 cc or ml. and the fossil from Israel had almost 1,750 cc or ml. In the case of 1,750 cc or ml the enlargement was 1/6 over the median 1,500 cc or ml; hence this case shown a life expectancy with almost 10% longer than the average. In fact an individual life span was 40-50 years, so the larger brain would extend the life expectancy to 44-55 years.

However this wisdom seemed to be known by the most of human cultures on Earth, and therefore the random practice to enlarge the skull after birth had an almost well-established scientific reasoning: to extend the life span and the intelligence with about 10%.

In the meantime, this wisdom led to the general idea that higher beings must have larger brains and skulls. The fossils found in South America had the tendency to enlarge the skull by 50%, while they expected 25% increases in life span and intelligence. We have no information on any results of this type of brain expansion, or on improved intelligence or on extended life span; hence we cannot comment any further on this particular subject.

***

Because we still talk on **brain development** I like to make a short analysis of the job being performed by the **brain's hemispheres and the implications on general thinking**.

The right hemisphere was the main part of the brain used before the development of language. It expressly participated in our "visual thinking", producing the analysis of spatial relations and their transformations. This hemisphere was analyzing the seen reality in a nonlinear way based on visual perception of "complex" and "real"; **it is like being-in-the-world, but where the world is complex**. The nonlinear analysis used simultaneous modes of processing and decided on "visual strategies" (different situations, opportunities were analyzed at once). Visual perception was cognitive activity and enlighten "balance". The pictographs were visual metaphors.

In nonlinear thinking the observer attempts to identify in the environment, the spatial frequency and/or the contrast produced by a stimulus exposed in the near or far field.

The appearance of symbols was the result that combined "visual-nonlinear" and "verbal-linear" thinking modes. The iconographic and ideographic art and writing still were relaying on "visual" thinking.

Our pre-history led to a turning point, where the "**verbal-linear thinking**" broke its balance with "**visual-nonlinear thinking**", and gradually begun to dominate in every single sense. Linear was characterized at its core by a very important feature of centrifugal motion, where we have central input and an output driven toward the exterior. This feature allowed "control" over the input and output and was called predictability.

By this feature **the scope of the system thinking is the "output" or the final product**. This approach is first recorded as the initial manmade diffusion of plants that led to plant domestication and agriculture. From this point on the "linear thinking" started to make progressive strides against the previous dominant "nonlinear".

Many scientists considered that "visual way" was using a **mental or mind map** to analyze the reality of the environment. In center of the map they placed the scope of their search (or the stimulus). From the environment came all possible interactions interfering with their scope. This system was similar with centripetal system of motion and the result of all inward interferences was transforming the original mental scope into a different product that was the "real-world-product". This **result took the form of "understanding"**, as opposed to computing, and expressed at least 2 million years of experience with such type of practice.

Since the linear turned dominant, the thinking mimicked the new type of neural connection that was shorter and direct, becoming the simplification named **cause to effect connection, or simple causality. It had gradually replaced the previous complex causality**.

***

The most important achievement of language was in the field of intercommunication and interaction; according to Hoffecker the minimum date for development of language was 50,000-60,000 years ago at the time of African dispersal. The language helped the Cro Magnons (who had the right mutations for speaking, in opposition to Neanderthals) to assemble together in larger groups and to assure a larger accumulation of knowledge by transmission from one generation to another.

The **settlements became the first centers for accumulative knowledge**, and they became entangled to other neighboring but similar centers. Within such settlements the chances to embed some super-smart individuals increases with the increased size of the settlement. Like in today world, those pre-historic and ancient super-smart were the true creators of the science and technology of that eras.

This accumulative process of knowledge became significate in amplitude when the practice of agriculture developed on larger scales, or in the last 8,000-4,500 years because the size and density of the settlements increased in a major way. This accumulation of population had produced, for the first time, an increase in knowledge on an extraordinary scale, comparing with any previous time in human evolution. 4,500 years ago appeared writing that was responsible for keeping a record of the acquired information. At this moment the linearity became prevalent.

This moment was the startup of human civilization (7,000-5,000 years ago) because of a large number of inhabitants that reached 5 million (8,000 BC) and 11 million (6,000 BC) people worldwide. It was a particular **numeric threshold (over 10 million people)** that was crossed 6,000 BC in Eurasia and northeast Africa. It allowed the creation of city-states and state/empire type of organization and produced significant areal densities.

However new linear thinking has brought many other major benefits, like control over the input-output and the prediction over the output. This allowed humans to make another major switching: they replaced the "power-with-hierarchies", which was copied from nature, with "power-over", or "command and control" hierarchies.

The previous "power-with" system considered intuitively that the universe is made of complexities and there exists systems within systems of increased complexity, and the complexity shows the real world situations. The participation and inclusion of all subjects was considered. This "power-with" approach was replaced by the "command and control" new approach as a direct result of growing inequalities and the access to power of the elites.

***

In the meantime another switching occurred: it referred to **hierarchical relationship**. This hierarchical concept can be define as "we look up for purpose and down for function". For example, a cell serves an organ and the organ looks down to the cell for function. The upper part of the hierarchy became the dominant elites, resulting from inequality, while the down part was taken by the suppressed, oppressed and abused, or by the poor.

During the Classical Antiquity it appeared the first non-branching-type-of-science. It was multidisciplinary. Understanding was contextual (as reminiscent of visual thinking) and was applied in various settings of multiple relationships, being able to bring enlightens over some complex systems problems.

During the evolution from Antiquity toward the middle ages, humanity turned toward **single-minded visions and understanding**. This single-minded approach focused on economic viability, which was pursued at the expense of environmental integrity and social justice. Linearity developed the spirit of competition that was opposed to the nonlinear spirit of cooperation and completion. In most cases, the competition was done against natural selection. Competition was manipulating to obtain short term success, producing in long run unsustainable development that was inherited by now for millennia.

Since the Hamitic and Semitic cultures of the Middle and Near East adopted the **linear features of the alphabet**, also embraced by Greeks and Romans, the **Western Hemisphere had slide increasingly more into the "linear thinking"**.

After the Renaissance (since 1250), Reformation, Commercial Revolution (since 1488) and Industrial Revolution (1760-1840) another shift away from experience occurred, leading to the rise of **representational thinking**, like the introduction of printed word (books) and the rise of the novel. All **industrial Western cultures are based on linear thinking, and so is their sciences**.

Only recently with the advent of film, electronics and computers, the images on the screens provide "visual models", but **at this time the "visual thinking" is emerging within linear cultural models**. This is so because thought is treated as an abstraction and because we separated the concept from percept.

***

All theories in science are linear in nature, including the computers (because they process in serial manner), the Quantum Mechanics and the Science of Complexity. The only nonlinear operation in quantum Mechanics is that of measurement. We cannot compute in nonlinear manner the equilibrium of density matrix of macroscopic ensembles because the resulting destructive interference between different states will eliminate the coherence. Quantum computation relates to coherence and decoherence of macroscopic ensembles.

In the meantime, in the last one hundred years of tremendous economic and scientific progress, the Eastern cultures remained partially connected to their traditional "visual thinking", preserving the iconographic writing and the "practical way of life". They try to maintain a tricognitive system (visual, verbal and music).

Some scientists think that the universe is a giant quantum computer, and consequently our brain and our DNA are types of quantum computers. Hence we have evolved in this quantum world into a complex system that gradually changed our brain into more and more complex design. The brain itself is a system within a system.

Quantum computing will deal with microscopic features, like independent atoms, electrons or ions and it is about their multiple states, which will collapse into one state that is the result.

By contrast, the "visual nonlinear thinking" (that is the visual coherence of cognition) works with multiple macroscopic elements of the environment, making them collapse toward the center (by a centripetal approach), where the stimulus is placed (causing its interference); the result of this analysis changes the characteristics of the stimulus (by interference that is a type of addition), making it complex, and provides the "**contextual understanding**" that is the answer to this entire nonlinear processing.

To conclude, in my opinion, what we call human intelligence only is the linear abilities developed in the last 7,000-5,000 years or less. At present most of the "contextual understanding" disappeared from our brain processing. Thus **we are left with an artificial type of intelligence that matches only the linear world we created**.

Here in the end, a small philosophical approach is needed. **Nature is governed (maybe in exclusivity) by many (or a group of) mathematical constants**, like e, pi, phi, and so on (there are very many like this). On the other hand, many other physical constants, like Planck (h), and the speed of light (c) are not universal.

All constants are irrational numbers, or even transcendental numbers (they cannot be written as a ratio of any two rational numbers), such as pi and e. For example, 22/7 is the approximation in use for them (that is approximately near the number 3). For phi (0.618) the approximation is 2/1 (as a rudimental 0.5), or 3/2 (as more elevated 0.66). **These constants**

are totally undefined in a dimensional-space, like we think of it (based on integer dimensions).

Which is the world, where they are "rational"? As it seems, they are a projection in our world that rationalizes them.

Our world rationalizes them (by the means of approximation) because it is an ARTIFICIALITY CREATED WORLD INSTEAD OF THE REAL NONLINEAR WORLD. They must make sense for our linear mind.

3. WHEN THE CIVILIZATION APPEARED?

Archeological records show that the first settlements appeared in Mesopotamia at Jericho (sometimes around 8,500 BC-7,300 BC and were made by people called Natufians) and at Lepenski Vir (9,500 BC-7,200 BC) in Serbia. In Anatolia was an old settlement at Catal Huyuk (6,300 BC-5,500 BC), but the temple and archeological site from Gobekli Tepe (Turkey), built by the hunter-gatherers, may indicate that **some sort of temporary settlements (unknown yet) have appeared 12,000 years ago in Anatolia, West Iran and Caucasus**. This may have marked the real beginning of human civilization, where nomadic people have built on-the-ground some permanent constructions for temples and temporary settlements just before the outbreak of agriculture.

However, there are evidences in different regions that some settlement started 25,000 years ago. The same evidence regards the use of symbols and some alleged artificial stone artifacts (Visoko, in Bosnia, Vinca/Starcevo Culture in West and North Balkans, continental shelf of East China Sea and some other places in the Persian Gulf). Practically from 25,000 years ago until 5,000 years ago (or for almost 20,000 years) there are evidences about some nonlinear strategy materialization by the use of linear tools. But none of these created civilization, or what was created was some discontinue shards.

It took a 20,000-30,000 years of long accumulation of mutations, and also it took several environmental forcing to activate these mutations accumulated inside a pool like feature.

Now Gobekli Tepe site contains maybe the oldest temple discovered yet, and up to latest knowledge on this matter, it has been built by hunter-gatherers of the Middle East. This is so because we do not have any indication on connecting settlements, and the temporary settlements were too weak to resist to the passage of time. There is interesting to see how a painted cave transformed into on-surface-built temple by nomadic people; this certainly resembles the idea of the painted caves (42,000-35,000 years ago) because many caves became, in these cases, the underground temples of the hunter-gatherers (nomads).

We will see later that theoretically the first agriculture and settlements seemed to have appeared on the exposed shelf of East China Sea around 15,000 years ago, and the first ceramic pots also appeared in China about 20,000 years ago, but shown a steady presence since 14,000 years ago.

It was a sharp change in human evolution that reflected much deeper changes in the brain of the "moderns". From gene comparison it results that since that time the synaptic transmission has significantly changed in human brain.

However the introduction of agriculture was important from the point of view that it introduced the practice of input-output relationship that later will favor the control, statehood, inequality and that was the first major linear achievement after the appearance of language.

From this moment on, or around 5,000 years ago, the linear thinking started building, within a very long row of sequences (step-after-step), the civilization as we know it. Since then everything humanity built or discovered was on the same linear path. Even then very few achievements, but quite major in significance, were made by the visual-contextual-nonlinear thinking. It is very hard today to separate the linear from nonlinear achievements because in the end they all look alike, and because the tools were linear.

So it is not the way it actually one thing looks like, but it is about the way it was conceived. And we have a plenty of examples, of all sort of things, done mysteriously spectacular from Neolithic to Antiquity and from megaliths, to pyramids and to very many other things, which all enter the category of "way (or well) ahead of time". They all show a nonlinear strategy.

There were individuals, who did all those things, and when they perished no one else knew how to replicate the same thing. **So it was not the local culture involved as a whole, but it was the case of single**

individuals. Very rarely such knowledge had been transmitted (mostly like a secret and hidden into some codes) to another individual, and when they were transmitted, this chain of knowledge was disrupted and discontinued within few generations.

In the meantime in the Eastern Culture it was a continual search for "context" that prevail over ages. In Hindu and Buddhist philosophies the meditation implied an "emptying of the mind" that in fact was removing the "linear thinking". It was used the right brain in the same manner as in contextual understanding and visual thinking.

In Daoist, or Taoist philosophy the practitioner was supposed to remove from his mind the "illusory and unreal structure of chaos" that was the linear society. The Early Chinese philosophy was characterized by non-systemic and non-step-by-step argument, while the contradictions were accepted. It was a reluctance to use words because they were mastering the daily linearity. It was here the same search for context and the acceptance for the meaning change as the context was changing.

However the Eastern Culture failed, too, to use the linear tools for nonlinear achievements because here was too much aversion for linear in all its forms, including the tools.

I would say our "context understanding" is now entirely gone in the Western Culture, and dangerously altered in the Eastern Culture, while the **artificial linear thinking, spread all over the planet, is non-intelligent from the natural point of view**. In short, we are today more inept to natural questions and answers than we were few thousand years ago.

So our civilization appeared, when linearity became the dominant strategy of human thinking that occurred some 7,000- 3,000 years ago, depending on location.

4. The Last Ice Age influence on human development.

The end of the Last Ice Age, marked by *deglaciation, has been a forcing environmental factor that influenced human development.* One can say that changes in the climate, and in the general conditions of the environment, have forced human brain to new adaptation. It has followed the trends appeared before the Last Glacial Maximum, while using these trending elements, new complexity arisen, producing new process-mechanism in the human brain.

Last Ice Age era lasted 100,000 years but in short it can be characterized by two very cold episodes: 75,000-58,000 years ago and 22,000-16,000 years ago (Last Glacial Maximum). Between these episodes (57,000-24,000 years ago, or for almost 30,000 years) it was an era of not very cold climate but stressed by small episodes (named interstadials) with duration of 2,000 years or so; the interstadials alternated from relatively cold and dry to mild warm and wet.

The *mildest episode of the Last Ice Age was 39,000-36,000 years ago.* At that time, most of the Cro-Magnons ("moderns") were already inside Europe for several thousands of years (probably 5,000 years). This is the time when we estimate that the Neanderthals begun to disappear. On the other hand this era had the optimum climate of the entire LIA, and it was a good moment to make "the moderns" cohabitate with local Neanderthals over newly defrosted lands.

41,000 years ago took place a temporary magnetic polarity reversal caused by 440 years long Laschamp geomagnetic excursion. As it appeared, the polarity reversal was 250 years long. The path of the excursion had crossed Pontic Basin and Anatolia, eventually exercising a local cosmogenic influence.

This event may be responsible for genetic mutations in both Cro-Magnons and Neanderthals. It is a very low possibility that the cosmogenic event of Laschamp (41,000 years ago) have caused some dangerous mutations in Neanderthals and Denisovans, influencing their demise.

A much more severe disturbance followed shortly, when 39,400 years ago Campanian Ignimbrite (located near Naples, Italy) eruption took place. It was a super-volcano eruption and the largest in the northern hemisphere in the last 100,000 years. It cooled local climate for several hundred years.

The peak of the low value of the geomagnetic field was between 40,000-30,000 years ago. Some scientists see the epochs of low magnetic intensity as the most favorable for human or homo development. Again this epoch strangely corresponds with the alleged Neanderthal demise. Or, we could say the **arguments against Neanderthal demise are very strong: the only other alternative would be that the Neanderthals became lost in interbreeding, becoming us**.

Next would be the mini-global-warming era (39,000-36,000 years ago); this would be the era when the "moderns" would be the most attracted to enter and move within Eurasian environment. This migration may have been quite large, adding many new comers to the already existing Cro-Magnons, while outnumbering the local Neanderthal population of the existing mixed communities; the ensuing evolution has caused the alleged fast Neanderthal demise.

However 30,780-28,140 years ago occurred another excursion, named **Lake Mungo**, when eventually new genetic mutations developed. It was followed 26,000 years ago by **another excursion**, and again 23,000 years ago by the **Mono Lake** excursion; this last one had triggered the Last Glacial Maximum (LGM) that started 22,000 years ago.

It is important to mention that the estimates on the amount of radiation (nuclides, and especially the 10 Be nuclide and C14 nuclide) reaching the ground, during the geomagnetic excursions, was at least double than normal. For example the estimate for Laschamp excursion shown an amount of radiation equal to 100 mSv.

Within the time of the excursion the geomagnetic field was 95% weaker than today, allowing a large amount of 10 Be nuclides production.

During deglaciation it occurred the Gothenburg excursion (from 14,230-13,690 years ago) that interrupted the deglaciation process, and it was followed by another excursion 12,000 years ago that triggered the immediate cooling of the Younger Dryas epoch (11,800-10,900 years ago).

All excursions were correlated in the same epoch with mountain glacier advances, meaning that they triggered episodes of cooling.

During the Last Ice Age, except the regions affected by glaciation, and the deserts affected by excessive aridity, the Eurasian climate was a little bit (2 centigrade lower than today average) colder than today climate.

For a good reason the *Siberia climate was significantly warmer and more humid that the European counterpart*: this good reason was the very **large glacial lake of the West Siberia**.

Solar radiation energy plays a much more decisive role in polar ecosystems than in temperate zones. The exposure to solar radiation of this lake had produced a significant change in local and surrounding regions climate. Also the big Siberian rivers carried a lot of heat to polar and sub-polar regions. Another important factor was the permanent light of the polar summer.

Two of the biggest rivers of Asia, Obi and Yenisei were draining into this lake because their normal outflow to the north was blocked from the Arctic Ocean by an ice dam. The lake itself was draining toward south into Lake Aral, and from there into Caspian Sea. Caspian Sea was draining into the Black Sea (crossing through Azov Sea), and further into the Mediterranean Sea. At that time this was the longest "river", or water way, of our planet (over 15.000 km long), stretching while flowing from Central Mongolia and Lake Baikal to Caspian, Black Sea, and Aegean Sea of the Mediterranean Sea, while reaching into the Atlantic Ocean at the Gibraltar Strait (that was like a long canyon in that time).

This hydrographic situation existed in various scenarios (with waterfalls or without them) from cca. 55,000 to 12,000 years ago, entering into the deglaciation era around 17,000 years ago and continued until 12,000 years ago when the West Siberia Lake disappeared by draining into the Arctic Ocean.

The West Siberian Lake affected Central Asia and Eastern Europe climate for almost half of the Last Ice Age duration, producing several climate refugees in Caucasus, Ukraine, eastern Anatolia and the southeastern piedmont of the Central Asia (including the foothills of Pamir and Altai mountains), Yenisei Basin and Lake Baikal Basin, and of course Beringia, where humans (and up to Altai the Neanderthals, too) were crowding. It turned today territory of Kazakhstan in frequent episodes of grassland, making it a corridor that connected Central Asia to the Eastern Europe.

For example a short description of a Russian climatologist on the LGM climate around Lake Baikal (Siberia) indicated that, in this particular time period, the winters were colder than today and the summers were almost the same warm as today.

However the main human circulation path has been from Yenisei Valley and Basin to Irtysh Valley to Ural River Valley, and from here toward Volga Valley, entering current Donbass Basin and Ukraine. A branching of this path was toward western Caspian shore and Caucasus. Even when the main waterway partially disappeared, after deglaciation (or 12,000-11,000 years ago), these corridors continued to function uninterrupted for all migration purposes, bringing people from Siberia to north Pontic Plains and to Caucasus, and in reverse, too.

According to genetic and archeological evidence, the same corridors were also used in reverse by the people from the Middle East and Europe toward the Western and Central Asia.

The last part of the Last Ice Age, or the epoch from 41,000 to 18,000 years ago, contributed substantially to the accumulation of mutations, which turned activated only in the last 7,000-5,000 years.

In the meantime this same epoch gave rise to new haplogroups (30,000 years ago the Mongoloids and 25,000 years ago the R1 haplogroup), which represent today the majority of people of this planet.

5. SELF-DOMESTICATION.

I consider that self-domestication started long time ago, and it **involved the creation of the group**. As the research discovered, there are genes which favor tameness and others which favor aggressiveness. The primordial group has involved the strict family members. Gradually the group has been extended to other members, who were not direct family. This extension has been made by artificial selection within the group by rejecting opposing behaviors, like aggressiveness, and adopting more social behaviors, like tameness. In fact **the group has evolved toward cooperation between members**. The selection within has driven away the aggressive members, who without collective help eventually perished due to environmental dangers.

The selection eliminated the linear competing behavior and developed a sense of cooperation that was essentially nonlinear. In long run the nonlinear of cooperation prevailed over linear of competition.

As we can observe within nature this process occurs among many biological systems, which enjoy the type of social group and display various levels of cooperation. From this point of view, the hominins have not been distinct of other species.

Also we know that many animals and biological systems use a sort of technology to achieve an easier feeding or other life goals. In this technology conjecture the species use stone, wood and other materials to improve their abilities.

However the **selection of the behavior of taming was not the single cause of self-domestication**, but it was the primordial constraint of selecting the group collaboration that led to other achievements. This process of behavioral selection has occurred during very many generations, and eventually was triggered by the local increasing of adverse conditions in the environment, which harsher conditions just favored group cooperation. This phenomenon was a regional process, and led to regionalism in hominin development.

Here it should be considered another element with influence on hominin development because it affected the birth process and the nature of the new born.

Again it probably **was a climate constraint that ultimately influenced the birth and the length of the childhood** and it manifested regionally at different times.

The self-educated groups were smarter because they were able to cooperate better with the adversities and other problems of the environment. Also they developed an accumulative cognition that was the sum cognition of the group.

Homo sapiens had an advantage compared with other primates: the **childhood was considerably longer**. This aspect along with tameness (selected within the group) encouraged further on the process of self-domestication, meaning that knowledge and practices were transmitted from parents to child. This process of education was able to further model the child behaviors and the resulted behavioral education was inherited by the next generation. **This gradually became an *accumulative process of self-domestication*, which was distinct in hominins compared with other primates or mammalians**.

New brain research shown that new neurons in adult hippocampus (specialized in memory and learning) are influenced by environmental factors and they serve to fine-tune to predict changes in environment. One third of hippocampus cells are renew during the lifetime.

Environmental influences have a profound effect on the adult brain in a wide range of mammalian species. The research has found that stressful experiences decrease the number of new neurons in hippocampus, and **sociability depends upon brain cells generated in adolescence**. A long adolescence, like in humans, can help develop considerably higher numbers of new neurons in hippocampus, producing larger brains with better social behaviors. When this production is interrupted in adolescence, the subject becomes extremely anti-social. So the events of the environment have influenced the process of self-domestication by increasing or decreasing sociability of the individuals and of the group. **A stress-modeled brain**

can facilitate adaptive responses to life in a stressful environment. When aversive experiences far outnumber rewarding ones, the system can reach a breaking point and produce a maladaptive outcome and cause the continue reduction in the production of new neurons.

As it seems to be the case, the **development of speaking** was the element that allowed a more efficient and direct form of **linear transmission of knowledge**. Here we had the case that new verbal education technique was linear at its origin. The scientific research indicates that there is co-evolution between language and manual praxis (production of tools). Linear production of tools (control of input-output) possibly has influenced the evolution of language.

However at this point in time, the previous **nonlinear survivability becomes gradually replaced by the new type of linear survivability**. Sharing knowledge and information by verbal expression, as new form of communication, generated an explosion in self-domestication among Homo sapiens communities. As a result the communities were able to enroll more members, to communicate (by visual and verbal means) with other communities and were capable to educate the children faster and more complex.

Hence self-domestication was influenced by certain stage reached in the evolution of speaking. The cave art, during the epoch 40,000-30,000 years ago, was an indication about the creation of larger communities and about the incipient use of speaking and music. It triggered an association between the spoken sound and the natural sounds. It incited toward the manipulation of sound.

The number of painted caves dramatically increased in the era 30,000-10,000 years ago, showing a steady progress in language and communication. During the same time interval, from Russia Far East to Urals were found over 100,000 stone art (this mostly covers Siberia) relicts.

Self-domestication ensured more humans living together, gradually selecting and developing new social skills. Here I want to specify that what we call natural selection was only a process similar to "resonance", where the interaction intermediated by resonance can amplify or annihilate. The result was a *"selection produced by amplification"*.

Self-domestication was in charge with amplifying abilities, and the speaking helped accumulate information in the form of education.

This produced a new type of intelligence: the "accumulated" type of intelligence, where knowledge and practices were transmitted verbally to new generations.

This accumulation of intelligence was the incipient form of "culture". Probably in the same epoch appeared the first symbols (a stone with symbols from Ningxia/China was dated 30,000 years old). Some of the precious knowledge acquired were inscribed on many petroglyphs, and which many were made around 25,000-15,000 years ago.

It is possible that "self-domestication" was the process influencing the relationship between Homo sapiens ("moderns") and Neanderthals, while causing rapid embedding of Neanderthals inside the communities of "moderns".

A recent paper on the study of Neanderthals found that their brain was similar to "moderns" at birth, being both elongated braincases. After birth (at the age of 1 or around) the Neanderthal child braincase did not turned into a globular shape, like in humans. More than that, the shape of the Neanderthal braincase was preserved the same for the whole life. The scientists considered this aspect as an indication that a differentiation from humans in *"the pattern and timing of the growth of the underlying brain circuitry"* (Philipp Gunz in Current Biology, November 9, 2010). Gunz also said: *"What our results suggest is that these genes might be linked with the speed and pattern of brain development"*. *"All interpretations about Neanderthal cognition will always be somewhat speculative"*.

Another study (Eliza Strickland, September 9, 2008, in the Proceedings of the National Academy of Science) was about the Neanderthal mothers. It concluded that *"the Neanderthal babies grew more quickly in their early years than Homo sapiens, which could have placed a burden on their mothers."* Holly Smith said: *"This energy intensive child rearing may have caused somewhat longer interbirth intervals, or somewhat older mothers"*.

And another study (published in Nature, November 2009) analyzes the role of FOXP2 gene with important role in speech and language. This gene exists in the brain inside basal ganglia and cerebellum. However it exists in a wide range of animals and birds, and besides the brain, it is also developed in lungs, esophagus and heart.

This new research (Nature) continues the study published in New Scientist in 2008 by Roger Highfield. The Nature paper came with the conclusion: *"the characteristic mutations that we see in human FOXP2 may indeed be more ancient than expected"* (that is *"between 200,000 to 100,000 years ago and matches the archeological estimates for the emergence of spoken language"*).

I would say that Neanderthals shown a somehow different shape of their brain that might imply a different wirering of the neuronal circuits. Also the childhood stage evolution for Neanderthals was different than for "moderns". Can it be somewhat compared to the autistic children of humans? Certainly the level of Neoteny was different in Neanderthals, and also it had an influence from self-domestication.

Last element to compare would be the birthing. As it appeared the Neanderthals had fewer children, hence their population was less dense (we even could say "random") that the population of "moderns" and their groups smaller, too. Neanderthal population was smaller and tended to stay in limited areas.

For example a Neanderthal child remains was estimated to be of 4 years of age. After its tooth analysis was completed it proved that the child was only 2 years old. Neanderthal children turned into adults at around the age of 14-15 years, or some 3-4 years earlier than the Cro-Magnon's children.

Neanderthals passed from Mousterian tools to the more advanced Chatelperronian tools at the time the Cro-Magnon entered Europe. During the same time (42,000 years ago) dated the first cave drawings in Nerja, Spain. There is proof that this cave art belonged to the Neanderthals. Creating paint from various materials and substances provides the faculty of understanding how materials can combine to form things with new properties.

Richard Klein at Stanford University made a theory about a genetic mutation around 41, 000 years ago that caused a sudden mental flowering. Nevertheless the art started in South Africa 77,000 years ago, and this event probably was connected with another specific mutation.

V.S. Ramachandran (UC-San Diego) proved that the **prehistoric symbols used in art originated into the magnification of stimuli**: the most salient part of the animals or personages they represent; the bison was blooms of meat, the bear was fat but powerfully jawed, the "Venuses", representing females, had exaggerated breasts and prominent genitalia.

Ramachandran *"speculates that our early ancestors were primed to respond to signs of health and successful reproduction"* (as Katy Waldman said in her article in The Big Questions, October 2012). The abstractions of the Paleolithic shown how homo mind processed the world.

The fact that the Neanderthals made cave art and jewelry it could be an indication that their brains processed the world not very distinct from the Cro Magnons.

Hence their inventively was not much behind the Cro-Magnons. But their culture was static, their art was rare and not so rich like that of Cro-Magnons. In other words **their creativity was limited by a low level of Neoteny.**

Another recent study of University of Oxford found that the **elongated regions in the back of Neanderthal's skulls corresponded to a larger cortex, but here was less space for the evolvement of the frontal lobes.** As a consequence the Neanderthals had an extremely evolved vision and visual thinking. But their "social brain" (frontal lobes) was smaller, meaning smaller social networks. It was calculated that the Neanderthal trade network was around 50 km, while the Cro-Magnon trading network was over 300 km.

The fact that Neanderthal children missed parietal and cerebellar development at an early age of their brain eventually generated problems for integrating sensory information and for abstract representation of the surroundings.

Another issue could be the way the Neanderthals risen their children, and how distinct was this from the Cro Magnon job on children. A limited Neoteny on Neanderthal part combined with poorer language abilities and distinct education of children probably made the greatest difference between these two types of homo, **placing the Neanderthals on the almost very slow path of evolution, which opposed the evolution of high dynamics shown by the Cro Magnons.**

However it was found that the Neanderthals were of three distinct types. For example the Amud Cave (Israel) specimens were around 60,000 years old and shown the largest brain ever found: 1,740 cubic centimeters (or ml.).

On the opposite side of Asia, in Guangdong Province (China), the Maba Man was a 126,000 years old specimen, showing a transition from

H. erectus to H. sapiens. This finding supports regional development theory, which speculates that **H. sapiens in Asia evolved out from Asian H. erectus and independent of the African development of H. sapiens**. This type, or another (H. sapiens migrating out of Africa), created 40,000 years ago the Asian type of H. sapiens. The self-domestication of the Asian type, along with genetic mutation gave rise to Mongoloids.

To conclude this chapter, I would say that the longest-childhood favored the self-domestication process that in turn had produced culture as the primer for civilization. The same important was the way these homo treat their children; when they treat them with education, the encephalization was positive; when they treat them with hard work only, the encephalization was negative.

The hunter-gatherers were nomads, so their entire life was direct survival with the lesson attached to it; their children had a direct participation at the entire life experience of the family and eventually assisted with some light work. Later the pastoralists and their children had a similar life experience.

6. Deglaciation.

The main change brought by deglaciation was less in the climate behavior but more in the characteristics of the landscape. The continental shelves become gradually inundated by the rising seas, killing many land bridges; the rivers gained tremendously increased flows, and humidity increased too, allowing new types of vegetal cover.

The change in environment (deglaciation) added some alternative more fit, causing humans to exploit the niche of opportunity.

The *change was gradual but in many respects it was dramatic*: some lands disappeared covered by waters, other lands turned defrost and ready for exploration and use, new vegetation cover was growing at higher rates, new animals were migrating within new lands. Also a vegetation migration has occurred, participating in changing the vegetal cover in general.

This dramatic change within the environment asked for new mental features, like much better communication skills between the individuals. In many places the landscape become unrecognizable, hence the humans have to collaborate better to overcome this and other new situations. New conditions shown by the environment have allowed to new paths of mobility, creating new itinerary of migration for the hunter-gatherers.

All these post-glacial novelties required new human abilities, especially in the domain of inter-human communication. The scientists discovered that the gene WBSCR17 was distinct in domestic dogs versus wolfs, while in humans it caused a syndrome called gregariousness and overly friendly behaviors. The same gene helped to read behavior and engage without restrains in social activities. It also increased tolerance toward one another. The humans became nicer and smarter.

This feature was gradually developed in humans during self-domestication epoch and allowed the spread of social behaviors and the better inter-communication that was assisted by verbal thinking. These new behaviors and abilities were manifesting in full swing at the end of glaciation and beginning of deglaciation.

However the self-domestication and domestication were not driven by a single gene but a group of them and **in the brain were no less than 14,000 genes**. This already mentioned gene (WBSCR17) had also some role in the higher reproduction rates. We learned that the trait, being selected, was causing nonlinear changes in some other traits, because between these genes were no causal relationships. The genes guiding the behavior, do so by altering neurochemicals in the brain.

The deglaciation era corresponded with the manifestation (maybe a later manifestation triggered by environmental pressure and forcing) of new genes and behaviors among humans. It also was the **era when the linearity of thinking was becoming prevalent**, while the new linear strategies were missing the big picture of environmental changes. Time and again, humans evolved regionally with specific time and rate differences between various regions.

In general the Nordic regions of Eurasia, where the main melting of glaciers occurred, were more affected by tremendous increases in the ground water volumes (larger rivers and lakes) and frequent riverine flooding. The regions on and near the continental shelves were affected by global ocean rise and flooding.

All these elements increased the mobility of human groups, which on the occasion were forced to move faster than usual toward other environments. This was evident when compared the distances travelled by humans from 60,000 to 40,000 years ago (so 20,000 years) versus the distances travelled during the time span of the main deglaciation events (which lasted 6,000-8,000 years in the continental Eurasia).

The increases in fertility were draw back by environmental catastrophes, like flooding, large avalanches, tsunamis which all were at their peak.

7. The mental sharp change.

The "moderns" become dominant by larger numbers (like 90% from total population), while the Neanderthals in this epoch (40,000 to 26,000 years ago) have been rapidly reduced to a mere 3-6% of the total combined population and rapidly absorbed. We know that Neanderthals use to live in small groups of 6-8 persons, which were self-sufficient, meaning they live in relative isolation, occupying the space randomly.

The Europeans show 3% (2.1% new estimate) of Neanderthal ancestry, while the East Asians show 6% of the same ancestry. Again Denisovans ancestry is higher in Asia (3-4%) than in Europe (under 2%).

Not all admixing from Neanderthals have survived in today population, so our estimates can be misleading, and the cohabitation of "moderns" and Neanderthals probably was much larger than the scientific data indicates.

The interbreeding was a direct advantage for the "moderns" (like the genetic mixing is known to produce into "moderns" the adaptation for skin, hair and immune system), also bringing to "moderns" a very long experience and adaptation with the Eurasian environment; but on the opposite, it may have suddenly turned the Neanderthals and Denisovans out of date and highly inefficient, and the ensuing competition was able to eliminate the rest of them extremely fast (in less than 5,000 years).

The successful interbreeding of "moderns" with Neanderthals occurred paternally on the side of Neanderthals. This may suggest that the **Neanderthal females were not capable of having children with "moderns"**. This data suggest that the **Neanderthal males dominated the mixed communities**; otherwise the Neanderthals as a whole will not be accepted in cohabitation. There is to suggest that also Neanderthals' male may had a role in hunting strategies of the mixed community because of extensive knowledge of the environment and long term experience with those places particular hunting habits.

Several studies focused on general population IQ of the countries of the world have suggested that higher IQ scores for Europe (R1a and R1b) can be associated with later adoption of farming (by R1a and R1b haplogroup). It shows that the populations (R1a and R1b) who practiced pastoralism until later in the Neolithic and early Antiquity (until 5,000 years ago in Central Europe) have produced strong cultures during Antiquity, while today they are still producing higher IQ (in Eastern Europe the pastoralists coexisted with farmers until 4,000 years ago).

The pastoralists invaded and conquered the farmer's cultures (which cultures ignited some 9,000-8,000 years ago) in an epoch around 3,500-1,500 BC, changing or widely altering local genetics. Most of the Antiquity achievements (from Sumer to Egypt, to India) started around 3,000 BC.

This information collaborates well with the idea that rapid verbal thinking of the farmers was developed detrimental to their general intelligence because it suddenly diminished the visual or spatial thinking.

To conclude, the sharpness consisted of:

- "**incipient speaking/verbal thinking manifestation**", the "**art manifestation**" and the "**musical manifestation**", all occurring around 42,000-35,000 years ago;

- **incipient culture** in the era 30,000-15,000 years ago that was the era associated with **self-domestication, incipient communities** and **accumulative intelligence**;

- the era 15,000-8,000 years ago, when **the process continued with domestication, settlement and diffusion**.

In only 30,000 years (in the interval 40,000-10,000 years ago) the "modern man" was built out to current complexity. But the main "modern" character of human brain appeared only in the last 8,000-4,500 years and it was entirely related to linear thinking and doing. The ancient linear society drawn mostly in the last 6,000-4,000 years is similar to contemporary linear society. The mental fundament of the society was made by a combination of skills, defining the visual analog representations and the verbal digital representations.

As it appears, the evolution of brain processing was marked by a split encountered by the populations of the Western culture versus the populations of the Eastern culture. The neurological research indicated that "faster reaction time to stimuli" of the eastern populations is connected to a more efficient neuronal system. In the meantime the eastern populations (especially in East Asia) have developed a considerably larger use of visual thinking compared to larger use of verbal thinking in the western populations.

The research on IQ tests indicated (Richard Lynn-Race Differences in Intelligence-2006) a **large differentiation between the principles, defining intelligence**, in the Western culture and the rest of the world (including Asia, Africa and Latin America). For example in the Eastern culture intelligence is viewed as means to recognize contradiction and complexity. For the Chinese, the intelligence means understanding and relating to others. It includes cleverness and responsibility. The Eastern Asians are visual learners, who want to see how something is done. Chinese and East Asian students in Western Universities are silent during class, and when they are forced to talk, this negatively influences their performance (being impaired by talking aloud when thinking), because in fact the East Asian silence means engagement in thinking, while verbal performance appears as a distraction from thinking.

All these differences shown that, because culture is experience, all IQ tests are culture bound, while **practical and academic intelligence develop independently, or even in conflict with each other**. Very basic psychological realities can be products of cultural beliefs that cannot be thought outside of their cultural contexts. Other studies have found a clear gap between the type of analysis performed in the aforementioned culture: **the westerners have a tendency to analyze the content (logically and verbally), while the easterners analyze the context (nonlinear, complex and visually)**. Again this shown a distinct type of neuronal processing between these two groups of population.

The study (of Richard Lynn) tried to find if this differentiation is due to environmental and/or genetic causes. It has been proposed that the genetic causes are the dominant ones, but the environment pressure had certain contribution, too.

Several neuroscience studies have been recently dedicated to differences in cognitive processes between Westerners and East Asians (on Chinese and Japanese subjects), caused by cultural biases to process information holistically (East Asians) or analytically (Westerners). The studies found **differences in cortical thickness and in gray matter densities**. The data suggest that experience and sustained biases in behavior, like those mediated by culture can influence and modify brain structure and/ or function. **Holistic bias to processing visual information results in larger hippocampal area in East Asians because of the cultural influence on "contextual" processing.**

It was found a **variation in brain shape**, where Japanese and Chinese have shorter and wider hemispheres and **increased tissue** in left frontal cortex and left middle temporal gyrus compared with Westerners. This can be caused by the requirements for fluency in spoken and written Chinese.

As it has been researched, the **East Asians and Westerners differ in cognitive processes because of "*collectivist*" versus "*individualistic*" biases endemic in East Asia and Western cultures, respectively.** As a result **East Asia tend to process information holistically and Westerners analytically.** Here can be added differences in diet, environmental exposure and genetics, which may have variable contributions.

As appears, the **differences tend to magnify with age because of experience and continuity of environmental factors.**

Another study on bipolar disorder found that the mental causes of this genetic disease are totally absent in the East Asians. Again, this is due to differences in brain processing.

An example on clear differentiation between Western and Eastern cultures is the **business of war**, which appeared around 4000 BC in Mesopotamia. There are enduring and ancient patterns in the way cultures approach war, because the strategy is rooted in timeless traditions. We have to consider the Middle East apart from West and East, but as a primordial and independent culture (Mesopotamia, Phoenicia, and Egypt) that gradually was conquered by the Western cultures (Greek and Roman), but also by local Arabic culture (for 1000 years), while culturally it became later more affiliated to the Western culture during the colonial and post-colonial eras; we can see a difference in philosophy of the West and East, where the inter-cultural boundary crosses somewhere east or west of the Indus Valley; this cultural differentiation led to different strategies in approaching and making the war.

Chinese statecraft and rulers were historically inspired by the **Confucius-Mencian concept** about the utility of force projection that was **non-expansionist, non-aggressive,** and preoccupied with internal anarchy; it was a prudent policy of **conciliation and compromise, being in tune to ancestral tradition of harmony.** They see that any war was never final and the danger of uprising might occur after the war had broken down many social and cultural barriers (and when the violence becomes daily business).

A dominant role had **Sun Tzu,** who had stressed the role of intelligence and deception, praised the ideal of **bloodless victory,** and **stressed the way of dialogue daring to protect the good standing of local economies,** and **finding a resolution without military ways.** But when the military action was inevitable, he recommended confusing maneuvers, surprise and deception. However the similar tactics were applied by the Aryans during their Eurasian conquest.

A similar concept was present in the rest of East Asia, in India and Southeast Asia, being somehow unified under the Buddhist traditions, but deeply and intimately based on local genetics. One can speculate that **Buddhism was ultimately a projection of the local genetics that spread from India to the East because it matched the existing genetics of the entire East and Southeast Asia.** Hinduism remained a local belief in India, where later the Buddhism was accepted along, or incorporated into Hinduism. This may show another genetic case, where Hinduism was in part an Aryan belief (that was a type of Zoroastrism), but in the form it was adopted in India matched the genetics of the locals that inherit Aryan genes. This type of genetic mixing is precisely expressed in Hinduism that appears as a transformation made to the Zoroastrian belief (this belief was in existence around 2000 BC) in order to match local conditions of the civilization from the Indus Valley (around 1500 BC).

However later migration of local civilization from Indus Valley to Ganga (Ganges) Valley (starting from 1000 BC), producing new genetic mixing, made the Hindu belief to evolve into the alternative of the Buddhism (500 BC). However, as appears, the individual genetics of Buddha were much closer to the East Asia (eventually Chino-Tibetan genetics) than local genetics. Hence the Buddhism spread first to Tibet and from there to China and the rest of East and Southeast Asia, and only in the end in the India proper.

This case reflects **a particular type of mental genes that allow particular ways of thinking, and which have influenced local beliefs of the Eastern Culture**. These genes favored the concept of totalitarian elites in Zoroastrian belief (Aryan genes), and then with the genetic mixing from the Indus Valley, the new form of belief, named Hinduism, favored the dominance of pastoral elites over the established society of farmers. Later the admixing of Sino-Tibetan genes favored the Buddhism. This last admixing of genes was promoting "contextual-visual thinking", compassion and harmony. **Buddhism was trying to heal the inequality promoted by the concept of castes and elites from the Hinduism and Zoroastrian beliefs**. Nevertheless, Hinduism was based on local contextual-visual thinking, but adopted the Aryan pastoralist concept of the elite dominance. Even then, for more than 3,000 years, the **Hindu society was based on democratic rules, as a special form of harmony within the inequality among castes**. By contrast, the Western Culture initiated social democracy only in the 18th Century AD.

A study from an international team led by Michael Bamshad (University of Utah) explored the impact of ancient Western migrations on people in India. The study said: *"Historians believe these West Eurasian immigrants established the present Hindu caste system, while appointing themselves to higher rank castes. Maternally inherited DNA was overall more similar to Asians than to Europeans, though similarity to Europeans increased with rank. Paternally inherited DNA was overall more similar to European than to Asians. The Western Eurasian immigrants mixing with native populations were mostly male, and that they tended to insert themselves into high ranking positions in the developing Hindu Indian caste system"*.

A paper written by Dr. Jim Sheedy comes with more arguments on Hindu belonging to the Eastern culture. He said: *"The Western Mind is oriented toward science, or the left brain analysis of the world around us. The West trusts science as the arbiter of truth about the world, thereby strongly anchoring the world view in the left brain. Another major difference between the Eastern and Western minds is their definition of success toward which the left brain strives. The West is driven by advancing the civilization. This is accomplished largely through science and technology. The Eastern Mind define success in terms of attaining harmony and enlightenment by approaching the truth about life."*

The paper emphasizes that the Eastern philosophy bears on the *"responsibilities of the individual toward the group and group harmony"*; in the meantime the Western Mind is described as being about the self or ego of the individual. The Western approach leads to hostile behaviors between individuals and stresses competition.

These utterly different *"approaches include profoundly different social relations, views about nature of the world. Differences are fundamental and can be clearly observed today-2,500 years after their formation"*.

Another differentiation between Western and Eastern cultures is about their perception on proportionality. Today in the West some 50% of adults cannot understand the meaning and the works of proportionality. In the East only 15-20% or even less show this inability. From Paleolithic to the late Neolithic the mind, that was nonlinear, shown proportionality in reasoning (cave and rock painting, various artifacts and constructs). This proportionality approach was generated by the context thinking that characterized the nonlinear.

Recent studies on fairness in cooperation indicate that the **proportionality was a basic morality in prehistoric times, and it begun to degenerate during the development of linear approaches, starting 7,000 years ago**.

However, pi and phi (irrational numbers) represent the proportion, or ratio between two quantities. We learn today that they exist almost in everything in nature, and are generative of order. More than that, the pi, phi and Fibonacci numbers are related by several mathematical methods, including in trigonometry. This is so because the cycles are the temporal aspect of circles; they are to time, what circles are to space. The curvature of space, the wave phenomena, the spirals, the rate of growth are other elements defined by the same pi and phi, and Fibonacci numbers. Many other constants also express fundamental, or governing rules of the Universe.

Finding Golden Ratio was an accident in Paleolithic and Neolithic practice, but later during the classic Antiquity it became an almost obsession in the search for perfection. By contrast pi was known 4,000 years ago by Babylonians, Vedic Aryans (India) and the Egyptians. In the meantime phi was known by heart in China, where it served the Traditional Chinese Medicine from at least 3,000 years ago. Always phi was discovered by accident in drawing proportions (ratios), so this may had occurred very far away in time.

I gave a quite extensive dimension to this issue because it defines a significate difference between individuals from the Western Culture against the Eastern Culture, while it defines human evolution in general. In the West it is a tendency to see the visual, contextual thinking as an avatar from Early Neolithic, or before the agriculture. To the opposite of this idea the **science proves that "visual, contextual, nonlinear thinking is very much alive in over 4 billion Asians and in their diaspora** (Chinese, Japanese, Korean, Mongolian, East Asians and Indians). This enter the category of sharp differentiation in human evolution that is the subject of this chapter.

Several researchers from 19th Century made an investigation in the Aryan Culture, based on scattered, while somehow poor information they had at the time, and advanced several observable characteristics of this culture. They suggested that Aryan society was communistic and highly democratic, as it was seen in the archeology of Aryan villages. Later they adopted a patriarchal society that became aristocratic and despotic while they migrated all over the Eurasia. Despotism came when the pastoralists took over the farmers and inserted the Patriarchal system. Before that moment the Aryan arable land was equally divided, but periodically redistributed to prevent unequal ownership (this tradition was preserved for centuries in Russia). Households were somehow isolated to preserve the full ownership over it. So only the houses shown ownership.

According with the aforementioned 19th Century investigation (1880), this type of Aryan civilization started around 6,000 BC, and gave rise to villages around 4,500 BC. It has been estimated that this development took place in the space from Ukraine to Caucasus and to Anatolia and Iran. From here it spread to Balkans and then back to Ukraine, Volga-Urals, moving to Western Siberia and Central Asia. They invented the Sanskrit language in Anatolia that was brought to India around 2,500 BC.

However, this early archeological investigation and its conclusions are not very far from our current conclusions.

Let's continue the discussion on characteristics of the Eastern Culture versus Western Culture.

The **signal on an ancestral lower violence in the Eastern Culture is provided, among other things, by the issue of slavery**. In China slavery was in and out many times, being banned for different intervals of time. However, the slavery manifested mostly like household servants (who in fact received a pay), was among war prisoners and convicts (who were punished with slavery). Sometimes people with too much debt sold themselves to clear the debt, or they sold one or two of their children as servants to the landlord or master. A somehow similar situation was in India, but here the outside conquerors (the Arabs and Mongols) took large numbers of slaves from local Indian population and traded them abroad. All household servants had a special relationship with their masters.

Such mild slavery type in the Eastern Culture was opposed to the extremely harsh situation in which the Western Culture slaves lived and worked.

We have the case of the **augmenting role of violence, as the societies became more linear in thinking and behaving**. The **adoption of the patriarchate** was in connection with the development of agriculture and was about the transmittal of wealth and property, and indirectly of power, to the successors. Simultaneously, war became the main state of affairs in the Western culture; it became institutionalized by the statehood as the essential instrument in dealing with political issues and economic situations.

Let's return to **human intelligence that is considered to be the main achievement of human evolution**. The measure of this ability takes the form of the IQ test. **The IQ test**, even being so controversial, can give us, in general, an idea about the evolution of human brain from Late Paleolithic to present. For example, the aboriginals of Tasmania (equivalent of Late Paleolithic culture) had an IQ score of 50 points, the Bushmen of South Africa (who are hunter-gatherers) 54 points, the Australian aboriginals 62 points, the New Guinea aboriginals 63 points and the Amazonia aboriginals from 50 points to 70 points. **On average these contemporary Stone Age people scored 50-70 points**. All these people are Stone Age people (equivalent to Mesolithic), excepting the Bushmen (equivalent to

Neolithic). However the Australian aboriginals show a variation from 52 points to 70 points.

Comparing the data from the above paragraph with the IQ of African population from different countries we have the following situation: Equatorial Guinea 59 points, Ghana 62-63 points, Nigeria 64-69 points, Sierra Leone 64 points, South Africa 63-77 points. As another example the Gypsies had scored 60-80 points. In this section the variation in IQ score is 59 to 80 points, which shows **an increasing of 10 points on average between the Stone Age populations and some regular populations living now on this planet**. Global IQ mean in 1950 was 91.64 points and in 2000 was 89.20 points. 25% of US current population has an IQ score of 70 to 90 points, and 50% score between 90 and 110 points. It is well researched the fact that children and adults who never were to school will score at least 10 points less than those who did. So, adding 10 points to our contemporary Stone Age people, their average score (70 to 90 points) equals now the score of regular populations from many countries (including 25% of US current population).

From 10,000 years ago to **1500 AD** (or in 9,500 years), different rates of development on different continents have produced a differentiation that placed Western Europe way ahead of **sub-Saharan Africa, Americas and Oceania, where all people were living in technological and socio-cultural conditions similar to Stone Age, Neolithic and Iron Age**. All these places were completely unknown to the Europeans before the 15th Century, but discovering and conquering them with the help of the incipient technology of the oncoming Industrial Revolution, the Europeans were able to colonize and control most of the planet, bringing for themselves an enormous amount of riches and prosperity.

The point here was to show that **our Mesolithic and Neolithic ancestors were no less intelligent than a large part of today population (like 25% of US population),** and our intelligence did not evolved significantly. Somehow in competition with this idea is the so called **Flynn Effect,** or model, that tries to solve the alleged IQ Paradox.

The Flynn's effect (Dickens-Flynn, 2001) is the explanation of the rise in mean IQ scores, which were recorded during the 20th Century. It shows that the **IQ affects the environment and the environment affects the IQ**, with the specification that this is the type of "short-live" effect of the environment. Two decades ago the saying was that humans influence the technology and the technology influences humans. It explains that **small environmental influences create large changes in IQ**, but these are, again, **short-live effects**. This system, being adopted from the Chaos Theory, still **allows heritability**. The **environment effect cannot accumulate over time, making a childhood IQ to differ from his/her adult IQ**. However a continual type of practice, mimicking the said effect, can make the change to last.

For our discussion on human intelligence, the Flynn effect has a particular significance. It is known that when species move to new environments, there are required increased cognitive demands for this new niche, and the species adapt.

This "effect" can explain fast adaptability to environmental changes during human migration that has been a bit of intelligence supplement to coop with encountered challenges idea of a short temporal effect is also very useful in dealing with short-time limited challenges. So, according to Flynn effect, the "barbarian invaders of Europe" were endowed with some supplemental intelligence.

We have **an extremely sharp change in our civilization course established in the last 500 years** by Renaissance, Reformism, Industrial Revolution, Colonization, World Wars, Decolonization, Communism, Social-democracy, Social and Wealth Inequality, Computers, Nuclear Energy, Nano-science, Genetics, Globalization, Information Technology and very many more scientific, financial, economic and politically significant events on a scale never experienced before in human history.

In the meantime, in the last 3-4 decades we witness a turn in the appreciation of the mental qualities and a quantification of these capacities by the IQ Test. Most of the literature associates the financial wealth with higher IQ scores. I think this issue is misleading because the highest level of wealth has been accumulated over many generation in a process that started at the end of the 18th Century and the beginning of the 19th Century. This type of wealth represents today over 99% of all wealth, and obviously this is in no connection to the past or current IQ performance, as we define it today.

Over the last 200 years the scientists made tremendous discoveries in all fields of science, which all brought **an enormous advantage to** the **Western Culture, and assured its worldwide dominance based on technology.** However this **technology proved to be an accelerating factor in the wealth accumulation and wealth inequality**, but the creators of this technology are only the employees of the Big Finances. They may be millionaires, or even billionaires, but in no way they control the Big Finance, or they have no word on any major decision making about the direction the technology and society should go.

Recently the startups become more and more numerous in the Western Culture, and more highly intelligent individuals are involve into it, but the main course remains under the control of the Big Finance. So the highest IQ bearers, as the creators of new technology, are of service to, and under direct guidance of the Big Finance.

For example 1 in 10,000 Americans may have an **IQ score of 160**, which at the level of US population represents 31,000 people, who are called the **super-smart**. From these people some 2/3 came for academics and for work from China, India, Russia and Eastern Europe. Probably, out of 31,000, only 10,000 are autochthones (30%), or American born. Even then **for every super-smart working in the US, there are 8 super-smart working in China and India (taking out the guest scientists, the real figure is 24 to 1)**. A similar situation occurs in the western countries of the European Union, where again the majority of scientists are from China, Russia, India and Eastern Europe.

This situation indicates that at present **the progress of the Western Culture, which is entirely based on science and technology, overwhelmingly depends on the best brains from the rest of the world. The rising of China, Russia and India, as major economic powers, will change the course of the world economics and finances**.

However the IQ is a quantitatively and linear approach that suggests that the **highly intelligent and super-intelligent people**, as based on their IQ performance, **will accelerate, by their contribution, the linear progress of human society**. More than that, the direction of the IQ products, and the technology they create, will be determined by the high financial interests, and it will not be for largely social benefits. This will also negatively affect the incipient nonlinear research and thinking.

Another characteristic of the linear thinking (in the Western culture) is "polarization", or the enhancement of the opposites. Here the high score IQ performers are situated at one pole, like everything else in the linear driven society. These high and the highest IQ scorers turn to be the opposite and the antagonists of the average and lower IQs, and therefore they contribute to increase the social differentiation and inequality. On the other hand, in the Eastern culture, the high IQ scores are seen in a different way, as previously described.

Visual Revolution (that is fundamentally nonlinear) **will come to conflict** (in the Western Culture) **to current acceleration of linear evolution that is said to be driven by the highest IQ performances**. This is so because the IQ was designed as a quantification of the linear (verbal, content driven, quantitative appreciation) abilities.

On the other hand the Eastern Culture is dominated by the visual (contextual, nonlinear, qualitative appreciation, and complex) thinking, and **any Visual Revolution will be more efficient and effective in the East then in the West**. This fact may drive human civilization to a significant split of achievement, leading to an increased contradiction between East and West.

Naturally the East has a mentally driven advantage over the West because of the differentiation in the thinking system that seems to have a genetic origin.

To conclude I would say that linearity had a complex origin, being it genetic, evolutionary and environmental. From 41,000 years ago to the beginning of current era (1 AD) the Eurasia was subject to 7 geomagnetic excursions, which totaled some 12,000 years, or 25% of this time interval. It has been assumed that many mutations took place in this interval in human brain, generating like a pool of mutations.

However, these mutations have been activated randomly by the forcing factors of the environment. There is more likely that some of these activations have been sharp and produced the tendency to practice the linearity. This practice was mostly inspired by the practice with language that manifested by extracting the subjects from their natural context and exposing them to the linear logics of analysis.

As we will see later in this text, after birth and until the age of 5, children still have 97% of their thinking expressed in nonlinear manner. This means, our change toward linear is a construct that increasingly develops in our life time.

In the meantime, the functioning of our brain in general is caused by the brain rhythms, while the rhythms dominating at the age of 3 to 5 years old are different that the rhythms occurring in our adulthood.

I would say the change toward linear thinking originated into a change in the brain's rhythms. And this change manifested sharp and recent. The physical proof is about the change in handedness. This change was occurring in the epoch 5,000-3,000 years ago. We will see more about it in the chapter Two Brains.

Another important change in human brain led to some changes into the sense of proportionality. As John A.J. Gowlett and other researchers have proved in their work, 1 million years ago hominins built bifaces, hand-axes of stone, respecting for a large to very large data set the value of 0.61 that is the Golden Ratio. It may happen to be the primary natural ratio in the stone flake, and it maybe be in our mind, too. So no one discovered the phi and phi because they always were in our right hemisphere.

Within particular ecosystemic communities, the proportions of species of different guilds are almost constant across the site in all stages of development. Prehistoric humans were part of this type of proportionality; numerous other proportionalities had abound in all ecosystems. **The sense of proportionality was inhabiting the inside and the outside worlds; it was into the mind and into the environment.**

8. THE COSMOGENIC FACTOR.

Wikipedia explains: "*The immediate dose from cosmic radiation is largely from muons, neutrons and electrons, and this dose varies in different parts of the world based largely on the geomagnetic field and altitude. This radiation is much more intense in the upper troposphere, around 10 km altitude. Cosmic rays also cause transmutation in the atmosphere, in which secondary radiation generated by the cosmic rays combines with atomic nuclei in the atmosphere to generate different nuclides. **These cosmogenic nuclides eventually reach the Earth's surface and can be incorporated into living organisms**. These nuclides varies slightly with short-term variation in cosmic ray flux*".

The flux of particles at the surface is dominated by high-energy muons, and other energy components that are less biologically effective. In normal circumstances muons contribute to 85% to the biological dose from cosmic rays. Enhanced muon flux directly affects living organisms. Globally annual average dose of radiation is 2.4 mSv/yr. and the cosmic component is 0.39 mSv/yr.

During lowering of the geomagnetic field (as during excursions) the cosmic component can rise at least to double the normal figures. Geological records from rock and ice analysis indicates some recent increases (in beryllium-10, or 10 Be) placed 68,000 years ago, 60,000 years ago, 34,000 years ago, 23,000 years ago and 12-11,000 years ago.

More recent research has found that cosmic ray flux is modulated not only by the geomagnetism, but also by the solar activity. During low solar activity, like the Maunder Minimum from 17th Century increased the cosmic ray flux by 33%. It was found a multiplier effect of minor solar variations, which generated significant variations within the atmosphere. Also changes in the geometry of the geomagnetic field (like tilting of dipole to lower latitudes) enhanced induction nucleation of clouds.

For example the gamma ray bursts, generated within the galaxy or beyond, are not recorded in the ice cores, but they generate, during the bursts, a plenty of radionuclides, which are incorporated in the living organisms and may be responsible for mutations.

Conversion of a signal from one form of energy to a signal in another form of energy is done by transducers, which can be natural or manmade. For example crystals convert electrical energy into ultrasound (mechanical energy). Electromagnetic radiation of any kind (gamma, X-rays, UV/visible, infrared) can give rise to **photoacoustic effect** that generates the formation of sound waves as a result of sound absorption (for example into water). Some of the generated ultrasound exhibits almost the same frequency spectrum as the input laser pulse (light pulse).

We have several types of absorption: light becoming sound (photoacoustic phenomenon), sound becoming light (sonoluminiscence), and absorbed electromagnetic and acoustic energies become heat, and produces chemical decomposition of substances. However we have sometimes a complex effect where the ultrasound, ozone and UV radiation are combined, and we can have **photoacoustic cavitation in water and/ or other fluids**.

Other types of transducers convert vibrations into magnetic field, or water and air pressure into magnetic field, or electrical signal. As another example, Transcranial Pulsed Ultrasound (TPU) of low frequency (exposure of 5.7 MHz) can manipulate excitable tissue and neuronal electrical properties of the cells in the brain and their synaptic plasticity.

The cosmogenic rays create radionuclides, which are in the air, water, plants, animals and people. For humans and animals it is an **effectiveness criteria for the transferring of radionuclides in the food chain**. For general population **the transfer in the Arctic is five times higher than in temperate areas because of the specific geomagnetic cut-offs of the Arctic**. For the population, consuming in large proportion, or in exclusivity, of natural food products, the transfer is more than 100 times higher than in temperate areas. The Arctic indigeneous people who rely heavily on terrestrial food products (like reindeer meat) have about 50 times higher exposure than non-indigeneous local population. Even in the Arctic it is a geographic variation in the land-based food chain deposition-lichen-reindeer-human.

The external exposure from natural radiation varies little over time and is about 0.85 mSv per year for cosmic ray and terrestrial gamma rays combined. Where natural terrestrial radiation is higher (like "hot" springs) the total dose can be higher. Natural radioactivity gives an internal dose via potassium-40 in the body.

For Arctic regions the average is 1.5 mSv per year, but may be higher in the areas of radon emanation from ground into dwelling.

Caribou/reindeer meat that has gathered naturally occurring radioactive polonium can also add to the load, giving a dose as high as 10 mSv per year. Local ecology governs the transfer from radionuclides in the food web that may contain freshwater fish, mushrooms, berries, fruits and other animal meats.

For Siberian taiga (forest) areas the radionuclides are known to populate the taiga canopy that create a radionuclide screening over the entire forest covered areas. In the zone **with ice complexes** (permafrost) there is a great deal of fossil burrows, which have the tendency to accumulate the radionuclides over the course of time, generating a **ground radiation level**, on average, of 2.0 mGy per year.

As we will see in the next paragraphs this radiation on the Paleolithic population of Siberia, living there for several thousands of years, had mostly a beneficial influence on their general health, especially when it was combined with local fish and marine mammals' diet.

Some area on Earth are subject to high dosage of natural radioactive radiation generated by the geological setting. Again Wikipedia explains: *"In the world in general, high natural background locales include Ramsar in Northern Iran, Guarapari in Brazil (a black beach), Karunagappalli in Kerala, India, Arkaroola in South Australia, Yangjiang in China (Guangdong Province)"* and in South Sudan.

So radioactive radiation can originate in cosmogenic rays, or it can have a geological cause here on Earth. However, there are certain factors, which contribute to a more or less strong penetration of cosmic rays to the ground surface, and this high level of radiation can have significant influence on biological organisms, and in our case study, on the human brain by causing mutations, and influencing human evolution.

Climatic effects of cloud formation induced by galactic cosmic rays (CRs) became recently a high interest because the variation in the

geomagnetic field intensity can change the climate by the CR flux modulation.

We know that changes in climate have been associated with planetary orbital factors and with changes in the geomagnetic field, namely the geomagnetic excursions. These "excursions" show a flipping of the magnetic axis from its regular north-south alignment toward new alignment, where temporary the magnetic poles become placed anywhere between 90 degrees and equator. Usually the flipping of the magnetic axis progresses until 45 degrees latitude before returning to regular 90 degrees position near the geographic poles.

The magnetic poles are the places where the geomagnetic field has the lowest intensity (hence it is termed a "magnetic anomaly"), allowing certain penetration of cosmogenic radiation to the ground. When the axis flipping occurs, these low intensity magnetic field areas move along new magnetic poles position, meaning anyplace between 90 degrees to equator, but most frequently between 75 and 40 degrees.

As it results, both high latitude areas are frequently subject to geomagnetic low intensity. Eurasia and North America, which contain the bulk of the planetary landmass, are mostly affected by these low geomagnetic intensity of the northern hemisphere; in the southern hemisphere the flipping affects the southern tips of South America and South Africa but also the southern Australia and Tasmania.

As long as the "moderns" have been inside Africa, they have encountered very rarely a cosmogenic influence, excepting the case of temporary geomagnetic reversal. But the Eurasians living north of 45 parallel, like H. erectus, Neanderthals and Denisovans, were frequently subject to these cosmogenic influences.

Recent paleomagnetic analysis has found evidence of cyclic abrupt changes to the geomagnetic field which correlate with volcanic eruptions in the region where the magnetic field inclination changes occurred, and also it was found that volcanic eruptions correlate to earthquakes and solar magnetic minimums.

Let's see the main geomagnetic excursions being recognized by most geophysicists. I will refer to those excursions defined for the last 80,000 years:

-Norwegian-Greenland Sea/Gaotai excursion was dated 70,000-80,000 years ago, but more recent investigations consider the date 55,000-66,000 years ago and even 59,000-63,000 years ago; according to old data it lasted almost 10,000 years; V. A. Bolshakov (2007) name it Hajimus and date it 80,000 years old; in Hexi Corridor (NW China) it was named **Gaotai event and dated 82,800-72,400 years ago**; the Gaotai event has produced the largest volcanic eruption in the last 2 million years, Mt. Toba eruption (74,000 years ago), and a severe cooling event around 70,000-65,000 years ago that was considered colder than the LGM; this excursion somehow incorporated two peaks in cosmogenic radiation 75,000 years ago and 63,000 years ago (Dr. Paul La Violette) (according with other sources, like Dr. G. M. Raisbeck, the 10 Be peaks were 68,000 and 60,000 years ago);

-Laschamp excursion is dated 41,000 years ago, and lasted some 400 years only; it generated Phlegraen Fields volcanic eruption (Naples-Italy) (39,400 years ago) and a sudden cooling (for several hundred years); it corresponds to formation of Asian general type of human;

-Mono Lake excursion dated 33,300-32,700 years ago, and again 32,200 to 31,500 years ago; it was sharply followed (or 2,000-3,000 years later) by the next excursion that is Lake Mungo;

-Lake Mungo excursion in China was determined at 28,400-25,800 years ago (in Hexi Corridor) and in the Arctic Ocean at 29,000-26,000 years ago;

-a less well determined excursion was recorded 26,000-23,000 years ago.

According with this information the geomagnetic field was in excursional mode for almost the entire interval 31,000-23,000 years ago (or 8,000 years) that matches well the 10 Be records from Greenland and Antarctic ice-cores (samples).

-Hilina Pali excursion (called also Lake Biwa/Japan and Lake Imuruk/Alaska) is dated around 18,000 years ago (found also in a core sample from Baffin Island-Canada); it corresponds to a climate instability at the beginning of deglaciation;

-Gothenburg excursion some 13,500-12,500 years ago; it triggered the cooling event of Younger Dryas (12,800-11,600 years ago); it is not clear, if this excursion has been repeated again around 11,000-10,000 years ago;

-Solovki excursion (7,500-4,500 years ago, or 3,000 years duration); according to other sources **this excursion had two significant flipping, which created two cooling events:** 8,200 years ago and 5,900 years ago, first one with a duration of 200-400 years, and the second one for 500-600 years; **this excursion corresponds with the setup of domestication of plants and animals and introduction of the linear thinking**; the most evidence of this excursion is in the Black and Caspian seas basins, in Caucasus, Anatolia and Mesopotamia; -Sterno-Etrussia excursion 2,800-2,200 years ago; corresponds with development of classic antiquity; according with other sources this excursion duration was 100-300 years long.

According to this counting **we had 9 geomagnetic excursions in less than 80,000 years, covering more than 30% of this time interval**. For the same time interval we had **3 extremely strong (catastrophic) volcanic eruptions, which have been connected to these excursions**:

-Mount Toba some 74,000 years ago that seemed to cause a bottleneck of human evolution in Africa and 1,000 years volcanic winter (it was connected to 10,000 years long Gaotai event);

-eruption from near Napoli some 39,400 years ago (it was connected to Laschamp event);

-Aira Caldera volcanic eruption in Kyushu some 28,000 years ago (it was connected to Lake Mungo geomagnetic event 30,000-28,000 years ago) that was followed by Hilina Pali excursion 26,000-23,000 years ago and the LGM event (22,000-20,000 years ago).

The first of three aforementioned events seems to be connected to a H. sapiens genetic bottleneck in Africa, to the coldest event of the Last Ice Age (70,000 years ago), to a genetic mutation that triggered the first expression of art in South Africa (70,000 years ago); it seems also connected to appearance of music and language (70,000-60,000 years ago), and it marked the beginning of self-domestication process among H. sapiens (60,000 years ago) that led to Cro-Magnons (45,000 years ago). This same event drove H. sapiens out of Africa (60,000 years ago).

The second event placed around Laschamp event (41,000 years ago) had been followed by volcanism, cooling and genetic mutations in H. sapiens.

The third event was formed by two connected geomagnetic excursions, was followed by volcanism, triggered the LGM, and also produced the mutations responsible for the modern haplogroups R1 and O3.

On top of all these major events we had the geomagnetic jerks with duration of several hundred years (200-400 years).

We will see later that the aforementioned flipping is a norm and quite frequently appears in the geomagnetic field. It is not a cycling, but a random magnetic phenomenon.

I would suggest that the first expression of art of South Africa, occurring 70,000 years ago, represented a very significant change in the processing that was developed in the "moderns' brain". It can be associated with a major geomagnetic flipping and/or alleged temporary reversal (Gaotai event) that occurred around that time and eventually produced mutations.

The mtDNA gene products play an important role in the energy metabolism of the cell, and a causative link between mtDNA mutations and cognitive ability is a very strong possibility. The brain is the most sensitive to partial bioenergetics defects or bioenergetics gains produced by mtDNA. Mutations in mtDNA are very common and constantly arising. Over 200 pathogenic nDNA mitochondrial mutation are known.

For example it was found that mitochondrial polymorphism can influence negatively the IQ of tested individual. After the person was treated for this issue, the IQ score risen 20 points.

The **low intensity occurrence that appears during geomagnetic excursions triggers cooling of the climate**. It is proof that many cooling episodes had been directly connected to this flipping of the geomagnetic field.

In the meantime the low geomagnetic intensity is a cut-off area, allowing higher penetration of cosmogenic radiation. In China for the last 2,000 years, the record of auroras was constantly kept. These auroras borealis (occurring in the northern hemisphere) are the luminous effect produced by the penetrating radiation. Usually they can be seen only closer to the poles or at high latitudes. During the flipping of the field the magnetic poles slide toward equator producing auroras at middle latitude, like in China. So all these Chinese recordings are a proof about the geomagnetic excursion occurrences in the last 2,000 years.

The **frequency of cosmogenic radiation, when reaching the ground, can interfere the working frequency of various genes, producing resonance, while augmenting their regular function or annihilating a previous function.** Cosmogenic radiation that usually is quite strong can be naturally switched to very weak energy and effects, matching the DNA weak-laser mode and other low frequencies of cells and genes.

Recent medical research in Korea has been able to disclose the Primo Vascular System (PVS) that is a system that seems to transport inside human body the energy incorporated by the photons, which enter the system through the electromagnetic sinks of the acupuncture points. The points on the skin are low electric resistance areas; the function of these points is similar with the role played by the cut-off sinks of geomagnetic intensity. These points on the skin receive photons from the ambient, emitting back photons from the body wasted (or not needed) energy.

In 2000 an experiment at University of California at Los Angeles validated the photon emission/reception in human subjects. However at that time the PVS was not yet re-discovered in South Korea, so no further research followed up.

As it is the case, human skin has an electric resistance enough high to protect the body against the penetration of ambient radiation. But in certain places the skin has gates (acupuncture points) to allow photon penetration. Further on, these photons enter PVS, from where their energy is transform into a compound named sanal and transported (in the form of oxygen) to the blood, lymph, all organs and brain.

All animals have various networks of PVS, while larger animals have larger networks. The plants use the photosynthesis for the same scope. Hence **solar and cosmogenic radiation affects, by the means of resonance, the entire biosphere**.

To conclude I have to say that the cosmogenic interference in human genetics has produced mutations characterized by a multitude of options, all ready to respond to various changes/forcing from the environment. **Most of the mutations were not activated for quite some time, while their sudden activation was triggered by sharp changes in the ambient** (this may be the case 70,000 years ago for humans from South Africa as a result of Gaotai event and Mount Toba eruption). As we will see later, the effect of natural nuclear radiation was studied in East India (Kerala) and the results shown an acceleration in mutation rate in local population, without producing any lethal or harsh effects on people.

A study on shielding and biological role of the Geomagnetic field (GMF), named Hypomagnetic field (HMF), had shown an acceleration in cell proliferation, when the shielding was largely removed. Lowering of HMF shielding had increased the neuronal cell division. The study concluded that human neuronal cells can respond to GMF shielding condition, but the production of any alleged disturbances in cognition had not been observed and not confirmed. It was observed only the alteration in neuronal cell cycle.

A MIT study (2012) on radiation influence on DNA found that the DNA damage occurs spontaneously even at background radiation levels, but most of the damage is fixed by DNA repair systems within each cell, in the end all damage appeared to have been repaired. It indicated that is a growing evidence that low doses of radiation are not as harmful as scientists thought. On the opposite many scientists think now that radiation hormosis model that chronic low level of ionizing radiation, in addition to background radiation, are beneficial by activating repair mechanism that protect against disease.

In the meantime it seems that the higher activity of the repair mechanism of DNA may be responsible to activate dormant mutations accumulated within a "pool of potential mutations", which have been inserted during the epochs of low intensity of the geomagnetic field (excursions) and higher cosmogenic radiation cycles. The activation of dormant mutations has been triggered by climate forcing associated with geomagnetic field behaviors.

On the other hand the repeating process of DNA repairs may have partially generated a new linear strategy (cause-effect) in the DNA functioning that was assimilated by the neuronal rhythms and brain functioning. Hence this linear strategy originated in the epoch 31,000-23,000 years ago. The excursions immediately following this epoch (which occurred 18,000 years ago, 13,500 years ago, 7,000-4,500 years ago and 2,800-2,200 years ago) have reinforced this cause-effect new strategy. **The main manifestation of new linear causality appeared recently, or during the Solovki excursion (7,500-4,500 years ago).**

However the Greek "logics" first manifested sometimes at the end of the Sterno- Etrussian excursion (800 BC-200 BC). From the same epoch dated Taoism, Confucianism, Buddhism, and Yoga.

At the end of this chapter I like to estimate the geomagnetic events influence on human development in the last 80,000 years.

As appears the Gaotai event (80,000-70,000 years ago) correlates with the first manifestation of art, and probably corresponded with the first manifestation of music (rhythm) and speech (70,000 years ago). It opened an era of self-domestication (from 50,000 years ago) and produced the main out of Africa migration (60,000-50,000 years ago). It may contained a major bottleneck in Africa (73,000 years ago), as a result of Mt. Toba catastrophic eruption 74,000 years ago.

The Laschamp event correlates well with the beginning of cave painting, more music (rhythm) and more speech abilities (41,000 years ago).

The intermittent-excursional era (31,000-23,000 years ago) seems responsible for inserting a new linear strategy in human mind. It corresponds to the first attempts toward agriculture (23,000 years ago in Mesopotamia and probably in the same era in East Asia).

Two major astrophysics events occurred 14,200 years ago (a supernova explosion that seemed to be responsible for generating alternating warming and cooling episodes from 14,500 to 12,750 years ago) and 12,750 years ago (a super solar flare that appeared to be responsible for generating the Younger Dryas cooling event at 12,750-11,500 years ago). Any significant mutations associated to these events?

Solovki excursion has eventually brought or activated mutation that produced (in the form of massive material realizations) the linear strategy (7,000-4,500). The end of this excursion corresponds with:

-the Sumerians came to Mesopotamia (5,300 years ago);

-it begun the building of the first city-states in Mesopotamia (the main four states built 5,000-4,500 years ago);

-it occurred massive advancement in agriculture (including irrigation);

-it take place the completion of all known domestications of plants and animals (all domesticated plants and animals, we know today, have been domesticated before 4,500 years ago).

-starts dynastic era in Egypt (around 4,950 years ago);

-the first of the 120-140 pyramids was built 4,630-4,611 years ago;

-it has produced the first but massive inequality and begun the first organized wars (4,500 years ago);

-it appeared the Linear A writing (in Greece) developed from previous Minoan writing symbols (or hieroglyphs) (4,500-4,000 years ago).

As we can see, it occurred a selective pressure that favored linear cognition. This last process manifested in full swing immediately after the Solovki excursion, or 4,500 years ago.

Now the Sterno-Etrussian geomagnetic excursion was also followed by extreme advancements (2,800-2,600/2,500 years ago, or **800 BC-600/500 BC**) in material realizations:

-classical Antiquity in Greece with its famous philosophers: Pythagoras (570-495 BC), Herodotus (484-425 BC), Socrates (469-399 BC), Aristophanes (446-385 BC), Plato (428-347 BC), Xenophon (430-354 BC), Archimedes (287-212 BC), and many others;

-Buddha Shakyamuni (born in Lumbini, Nepal, in approximately 624 BC) who is the founder of Buddhism;

-Lao Tzu (born around 580-560 BC, on Yangtze Valley, west of Shanghai) who wrote Tao-te Ching and is the founder of Taoist philosophy;

-Confucius (551-479 BC) the founder of Confucianism.

Maybe here is an amazing coincidence about the influence of this excursion over the northern hemisphere and human achievements.

A low incidence was always given to the **Medieval Chill** that manifested between 1,250-1,300 and 1850 AD (it included the Maunder Minimum from 1620 to 1730 AD and Dalton Minimum from 1790 to 1830 AD), and which **increased the cosmogenic ray flux by 33%**. In a way or another, this chilling period of the climate, that was more severe in the Western Europe and North America than in the rest of northern hemisphere, corresponds with cosmogenic radiation increases, which seem to be responsible for the advent of Renaissance, Black Plague, creation of banking and financial system, Protestant Reformation (of church), European exploration and colonization of the world, Industrial Revolution, urbanization, creation of capitalism and massive social changes.

Because the exact local data are still poor, it is hard to connect the Mongolian invasion, Turkic and Arab conquests with clear climate and cosmogenic forcing. Even then some connections are obvious.

This mixing of natural and human events had an extreme influence on the evolution of human civilization, enhancing the linear strategy of thinking.

9. Music-rhythm in mental development.

As Howard Gardner put it, **it was the music-rhythm that was able to solve many other problems in the brain in novel ways.**

This new mental ability generated a cultural revolution, producing a variety of artifacts: amulets, small sculptures, jewelry, etc. However the archeologists indicate that a similar phenomenon (70,000 years ago) was present at that time in Europe, too, affecting on a different scale the exclusive population of Neanderthals of this continent. The revolution was not limited to cultural aspects, but had outcomes in tools processing and other domains.

The second eruption of art, that was represented by the cave painting, has appeared around 42,000 years ago in the Eastern Europe (Romania, in Apuseni Mountains), spreading to other climate refugees represented by the caves from France to Spain (40,000 to 38,000 years ago).

The evolution of art was partially constrained and obstructed by the harsh environmental conditions occurring around 22-18,000 years ago during the LGM (Last Glacial Maximum). The deglaciation, starting 16,000-15,000 years ago, has allowed the benefits of newly developed "verbal thinking" to be expressed in new domains, like in field orientation/ trip itinerary, hunting coordination, working strategies, communitarian gathering, collective activities, construction of tools and settlements, sharing experience and emotions.

Emily Saarman reports, (May 31, 2006) on Symposium, explores the therapeutic effects of rhythmic music: "*There is a growing body of neuroscientists who support the theory that if there's a physical of conscious experience, it has to be in the brain waves. It seems to be the only thing* *in your head that changes rapidly enough to explain real-time changes in consciousness" said Turow". "**Music with strong beat stimulates the brain and ultimately causes brainwaves to resonate** in time with the rhythm, research has shown. **Slow beats encourage the slow brainwaves that are associated with hypnotic or meditative states**. Faster beats may encourage more alert and concentrated thinking. Studies of rhythms and the brain have shown that **a combination of rhythmic light and sound simulation has the greatest effect on brainwave frequency**, although sound alone can change brain activity."*

Another conclusion at the Symposium was that rhythm, light and sound improve cognition performance by increasing the blood flow in the brain. Also tonal sound with frequencies range from 90 to 120 MHz induce special brain states similar to meditation. Recently a great deal of research have occurred in the new domain of archeoacoustics.

Developing a neuronal circuitry for speech was more important that morphological changes to the skull and other body's area contributing to sound modulation. This shows again that some genetic changes may have occurred long before they were able to support certain skills. The same case occurred for Denisovans, who gain high altitude adaptation, while the beneficiary were the resultants of their mixing with Tibetans.

The first form of verbal expression was probably a symbolically mediated language, where the symbols resulted from visual or spatial thinking. It was associated to a cognitive basis for symbol representation.

In all these, the sound of music (probably initially only in the form of rhythm) has produced the first gathering of the group of hunters. Such gathering may incited to incipient singing/praying and incipient verbal communication.

Probably the first gathering took place in caves and the added painting on the deeper walls required the torch lights to illuminate them, making the group to complete the solemn scene with rhythm and incantation. Rhythm was the frequency that resonated the brain neurons, eventually causing special states of mind (some type of hallucination). These special states turned supported and deepened with the ingestion of particular plants. Everything turned into a praying for good luck in hunting. Gradually such cave processions turned more and more wordy.

Here I can conclude that *cave painting of Europe is an indirect indication that "verbal thinking" has been on the track of fast development*. Singing or incantation in caves has been a form of "verbal thinking", and it may suggest gradual widening of verbal communication between the cave dwellers.

In the meantime the cave painting shows that the dwellers have reached *certain level of abstraction in thinking* that is confirmed by the other art expressions performed in the same era and at the same time. It is scientific proof **that music helps create and understand abstractions.**

Current limited archeological record do not allow to extrapolate the European case to other parts of the world, like Asia. Even then, we can speculate that similar processes have occurred in many places of Asia.

To summarize, we have two new elements in "moderns' brain" processing, which came almost in the same time: music-rhythm and verbal thinking.

10. Verbal Thinking.

The issue of verbal thinking and language is still a matter of large scientific debate. Recently it has been on Edge Foundation site a round table debate generated by V.S. Ramachandran paper on *mirror neurons and imitating learning*. The author supports the idea that mirror neurons have generated a Big Bang in human development about 40,000 years ago. The round table discussions have put together famous scientists, who have long argued on pro and contra issues but in the end no reliable conclusion has been reached.

Giving to the 40,000 years old Big Bang issue a second thought, it has turned to my attention an ignored fact: not long ago various researchers from Russia to Germany have discovered that our DNA is conversational because our language has a grammar and syntax similar to DNA own processing mode. If this is the case, what we need for the startup of language is a way to connect our brain with the processing mode of the DNA. What can cause such a connection? It must be **a gene mutation capable to open the way to read our DNA operating modus,** and then all previously accumulated skills can assemble together to produce, somehow at sudden, a spoken language and its needed circuitry in the brain.

Taking about the mirror neurons it seems, obvious that they may have an important role in learning and in accumulation of learning.

Like in every other case of human development this event of language has occurred at different times in different places, but as it may appear, the common cause, that triggered it, has been Laschamp geomagnetic excursion of 41,000 years ago. In an early stage the acoustic rhythmization (drumming, chest drumming and the kind) served in hominid groups for the purpose of synchronization. It is rhythm, consonance, coherence, balance, symmetry, pleasing proportions, equilibrium and harmony, which all together combined contributed to the emergence of music and language.

Language has many forms of symmetry because it is multilayered and hierarchical. Here language and music have in common the hierarchic structure, like elements-words-sentences-phrases-narratives-compositions. In the same hierarchic mode occurs mental problem solving and the using of instruments and tools.

The inventory of sound in a language tend to be symmetrical. Language is constructed from analogous frames, which are logical frameworks, or symmetrical connections. Also a general symmetry manifests between the nominal domain and verbal domain. For there to be meaning, the language has to use the codifying of the perception of reality. However, the language operates according to rules, and rules mean prediction that makes language a linear that also provides a linear structure. Being a multilayered approach, the multiform total symmetry brings linearity.

Let's see what advantages verbal thinking brought to the brain processing.

First, verbal thinking has introduced *"linear thinking"* that has made a short-cut in mental processing. It was a new step that brought superficial efficiency and rapidity in judgment (but it brought the cause-effect evaluation). Secondly it allowed smaller space for storage in the brain and tremendously improved the capacity of communication between individuals.

Here are some other radical changes:

-it helped processing of wordy abstractions;
-it allowed changes toward lesser energetic diets (because the brain needed less energy);
-hunter-gatherers gained some free time (saved time), allowing them to spear time for producing art and building better temporary or semi-permanent settlements;
-visual thinking was to male advantage; verbal thinking was to female advantage; combining both it produced a win-win approach, but mostly it introduced a better female efficiency.

However the development of verbal thinking, starting 40,000-20,000 years ago, caused later (starting from 20,000 years ago, but especially from 10,000 years ago) a reduction in endocrinal volume, which reduction is evaluated today at 10%. In women the change has been more dramatic, diminishing the brain from an average 1502 ml (30,000 years ago) to only 1241 ml today that is almost 20%.

Verbal intelligence helped develop the "cumulative intelligence" because it allowed transmission of knowledge from one generation to another.

Verbal intelligence was nicely combined with rhythm-music evolving abilities.

11. GENETIC CHANGES THAT CREATED NEW TYPES OF "MODERNS".

Around the same era some genetic changes have occurred among the "moderns". Two of them seem to be of undoubtable importance: the Mal'ta boy (24,000 years ago) in Southern Siberia and the first mongoloids (cca. 30,000 years ago) of Southeast Asia and/or South China. Both genetic changes occurred just before the Last Glacial Maximum. They represent the split between general Asian and East Asian traits that occurred 30,000-20,000 years ago.

However the Mal'ta boy is the oldest anatomically modern human reported to date. This site and Afontova Gora2 (on western shore of Yenisei River, some hundreds kilometers to the west of Mal'ta) dated 17, 000 years ago and the same MA-1 haplogroup, indicate that the entire South Siberia has a genetic continuity during the LGM.

Mal'ta boy has another implication, too: his mother has been haplogroup U, which is common for Cro Magnons. This means the Cro-Magnons diffused from the Middle East, or from West Asia, or Eastern Europe, toward the Lake Baikal (around 25,000 years ago) just after their early diffusing into Europe, where they arrived 45-43,000 years ago. Thus around Lake Baikal we have the mixing of "moderns", who migrated, from the Middle East, and/or West Eurasia, into South Siberia.

The genetic analysts consider that these changes are the outbreak that has produced the most modern humans. Both changes suggest an apparent adaptation of local "moderns" to arctic and/or subarctic conditions.

Global population studies have shown that there are striking differences in the nature of the mtDNA, namely because only two mtDNA lineages, haplogroups M and N left northeast Africa to colonize Eurasia. A striking discontinuity in the frequency of haplogroups A, C, D and G between central Asia and Siberia, where the environmental factors enriched for certain mtDNA lineage, when humans toward northern latitudes and had an adaptation to colder climates. Hence the genes of mtDNA are central to energy production the most needed in the cold.

Haplogroup R1a is believed to have originated in Eurasian Forest Steppes north of Black and Caspian seas. This lineage originated in the populations of Kurgan cultures. Haplogroup R1a, defined by the SNP marker SRY 10831B is part of the ancestral R1-M173 of which ancestors first arrived in Europe from West Asia in the epoch 35,000-40,000 years ago. During LGM people with Haplogroup R1a Y-chromosomes retreated in Europe below the ice sheets of that era, but clustered in the refugees placed north of Black and Caspian Seas (named Pontic plains, or the Forest-Steppe). After the end of the LGM, during the deglaciation R1a migrated north and eastward.

Alan Cooper and Wolfgang Haak in The Conversation (March 23, 2015), referring to a very recent study done by Harvard University, explain: "*Instead of sequencing the entire genome from a very small number of well-preserved skeletons, we analyzed 400,000 small genetic markers right across the genome. This made possible to rapidly survey large numbers of skeletons from all across Europe and Eurasia. Our survey shown that skeletons of the* **Yamna culture** *from Russian/Ukrainian grasslands north of Black sea, buried in large mounds known as kurgans, turned out to be the genetic source we were missing. The group of pastoralists, with domestic horses and oxen-drawn wheeled carts, appear to be responsible for up to 75% of the genomic DNA seen in Central European cultures 4,500 years ago, known as Corded Ware Culture. This must represented a major wave of people, along with all their cultural and technological baggage.*" These people were the haplogroup R 1a.

From haplogroup R1a, around 16,000-14,000 years ago, split the haplogroup R1b (also called the Arbins) that expanded throughout Europe around 10,000-12,000 years ago, reaching and settling mostly in the Western Europe.

The DNA mutation at Mal'ta (near Lake Baikal in Siberia), producing R1 haplogroup (26,000 years ago), changed the cell metabolism and produced higher cognitive ability that is today inherited by all Europeans bearers of R1a and R1b haplogroups.

The haplogroup R1 evolved inside a territory exposed to subarctic climate conditions. Another very recent study has found that the Europeans (R1a and R1b haplogroups) have a higher mutation rate due to a sensibility to UV (ultraviolet) light. The same UV (or the non-damaging part of it) is the main radiation domain used by DNA for communication and interaction.

However the mongoloids (haplogroup O3-M122) appeared first in subtropical South China; hence here were no arctic/subarctic conditions to adapt to. In our concept, the Mongoloids match a classic example of arctic/subarctic (or cold) adaptation. At present we have Mongoloid populations east of Lake Baikal, but they originated in subtropical South China from where they migrated to Siberia and Americas.

Neoteny can develop in environments were local conditions can inhibit the completion of metamorphosis, like the lack of available iodine retard the action of thyroid gland, or the same result can arise in the arctic conditions. Probably the Mongoloids of South China qualified for the lack of iodine salt in their food that produced their Neoteny.

Iodine deficiency is geological and characteristic for soils of various regions of the world (118 countries are listed for this deficiency within particular regions), including the coastal regions. The deficiency comes with the ingestion of plants grew on such soils (the result is mental retardation or cretinism). Undoubtedly the process can be inverted, when care is taken on time.

Several studies were made in Guangdong Province, in Yangjiang area where it is a high background radiation that is of geologic essence, and is three times higher than average. This background radiation has an influence over the thyroid gland in women. The studies measured serum thyroid hormone level, urinary iodine concentrations, and chromosome aberrations. Women in high background region had significantly lower concentration of urinary iodine and significant higher frequencies of stable and unstable chromosome aberrations. Increased intake of allium vegetable (garlic and onion) decreased the risk of nodular disease. The conclusion was that increased continuous exposure to low level radiation throughout life did not increase the risk of thyroid cancer. I would insist that this connection with iodine eventually was responsible in the first place for Neoteny; also this area may be at the origin of the first born individual with Mongoloid characteristics.

However the original process was complex and shown several possible scenarios, like the case of the local iodine salt deficiency, occurring before and during pregnancy, but which was inverted at some point, preventing any damages to the brain of the infant and pregnant mother. The inverting probably was sharp and took the form of adding iodine salt and a fish diet until the mutation toward Neoteny occurred (new gene SRGAC2) without damages. This seems like a local cultural tradition that banned the iodine salt for a certain time out of pregnancy diet, or forcing pregnant woman into some grain diet that lacked iodine salt.

The scientific discussion on Mongoloids mainly indicates that we deal with a pedomorphic development: the Mongoloids were deeply fetalized and thus capable of the greatest development. Stephen Jay Gould characterizes the **Mongoloids as Neotenic**, where an infantile or child-like body is preserve in adult life. This explain the development of disproportionally large brain.

Neoteny is characterized by the development of new gene SRGAC2 that slow spine maturation, increasing neuronal migration, increases the dendrite spines, which become more complex. It causes more synaptic densities. It produces a significant excess of genes related to development of prefrontal cortex. In the end it translates in adults in higher ability to learn.

As it seems the entire process, leading to Neoteny, was a controlled one, following a voluntary or involuntary process; such a cultural trend might was in place in the epoch 30,000-10,000 years ago in the South China (probably somewhere around the Taiwanese coast).

(Similar characteristics of Neoteny develop in domesticated animals, and their reason for Neoteny is the presence around them of **caregivers**. Caregiving to an entire community by the community itself can produce self-domestication as the whole or only to some violent inclined individuals. However it exists unproved ideas that human caregiving greatly helped the process of plant domestication.)

The archeologists shown that around 45,000 years ago occurred the separation of Asian and European types; 40,000 years ago it appeared an *Asian generalized cranial morphology (Tianyuan Cave, near Beijing, where some inherited robust Neanderthal features were present in the Asian type of skeleton)*, but between 20,000 and 12,000 years ago this generalized type disappeared, while new Mongoloid type emerged. Archeological records find that this *clear Mongoloid type exists only since 10,000 years ago or since Holocene.*

The Mongoloid (haplogroup O3-M122 that is predominant 45% among the Han Chinese) traits appeared in tropical China (25,000-30,000 years ago), and **this Neotenic trait became later the best adaptation for arctic/subarctic inhabitation**. This helped the Mongoloids to successfully colonize all arctic and subarctic regions of the Nordic hemisphere (starting 25,000 years ago).

In the case of Denisovan, the bones analyzed were found in Denisova Cave of Altai Mountains that is at 640 m altitude (this is a hilly altitude not an alpine one). But the Denisovans shown an adaptation to high altitude.

In the meantime, Denisova Cave is located in the Altai Mountains, with peaks, which tower to more than 4,000 m altitude. But no one has yet the proof that Denisovans were frequently climbing and camping at 3,500-4,000 m altitude. Probably again we have the case that a particular adaptation can cover several distant domains. This indicates the complexity of adaptation that is the complexity of the self-organization response.

Somehow the R1 haplogroup (including R1a and R1b) seem to enter a similar category. They are said to show subarctic adaptation. Nevertheless their adaptation (in physical traits) is strikingly opposed to that of Mongoloids. But they undoubtedly originated within the arctic/subarctic environments of Siberia (mostly in Baikal-Angara-Yenisei-climate refuges).

These two new haplogroups (R1 and the Mongoloids) form today the basal genetics of the majority of the world population, and **these people are no older than 20,000-10,000 years**. However the **true racial characteristics of these two haplogroups became evident only around Holocene (8,000-6,000 years ago)**.

The scientific analysis of mutation tends to be linear (cause-effect), and thus it may mislead us about the nonlinear outcome of this process. A mutation may bear several options (which are all hidden to linear analysis) and it may not be an exact outcome of certain forcing because it is part of self-organization process.

Thus we may never have an exact adaptation to a particular cause because the adaptation must be nonlinear, covering multiple options in the same time.

12. DOMESTICATION.

It is said that **Broad Spectrum Revolution pre-dated agriculture and appeared around 23,000 years ago, and produced a pre-agriculture adaptation**. It occurred multiple temporary domestications. Small seeded grasses and legumes occurred in abundance in the Upper Paleolithic in many places from Mesopotamia to Anatolia and to Caucasus, to India and to China. It was partially disrupted or altered during the Younger Dryas (12,800-11,600 years) but otherwise it existed during the entire deglaciation epoch starting 18,000-17,000 years ago. After the cold episode of the Younger Dryas (after 11,500 years ago) the domestication became relatively widespread in many regions of Eurasia.

The development of domestication of animals and plants occurred for many generations who were able to teach the subsequent generations to perform this task of domestication, until it became a behavior that was very linear in essence. It was a trial and error, and the accumulation of knowledge was handed down by language, so it was externally and not by DNA. However this was the beginning of the biggest suppression process applied ever to human brain: the natural nonlinear turned suppressed by new linear thinking and behaviors. This accumulation of linearly thinking and behaving people, or these farmers, taken together, they acted as one single huge linearly driven brain. This huge social brain of the farmers was spreading and constructing further more and more linear rules.

I like to quote Dr. Gil Stein (2010), the director of Oriental Institute of the University of Chicago in a paper on Global Perspective: … *"there is no single "center" of civilization from which all cultural development emanated. Instead, people in different parts of the world seem to have made similar types of major cultural developments (e.g. domestication and urbanism) independently, and always in ways unique and specific to that particular culture."*

"Hunter-gatherers practice strict population control to keep their numbers in balance with scarce resources and with their need to be highly mobile. When they settle, their population immediately starts to rise as all or most of the practices to keep birth rate low are relaxed. Birth spacing becomes shorter, and nutritional levels get higher". *"Sedentary hunter-gatherers focus on narrow range of very rich food resources (running a high risk of starvation)".* *"Hunter-gatherers need to store food since they can no longer move around the landscape to forage for spatially scattered resources".* *"Sedentary hunter-gatherers tend to over-exploit their immediate territory".* *"Sedentary hunter-gatherers can accumulate property."*

Some current estimates indicate that 22,000-13,000 years ago the population of Europe (excluding Iberia) was 9,000-40,000 people and the world population 12,000-13,000 years ago was 4-6 million people. After the beginning of agriculture, or around 6,000 years ago, the world population was 60-70 million people.

Also according to Dr. Gil Stein: first plant domestication in the Middle East was 8300 BCE, and first animal domestication in the same area was 7500 BCE. According to this data the first settlements predated these domestication achievements.

But according to the last archeological discoveries, the **pre-agriculture use of stone tools to process seeds appeared in the Fertile Crescent around 23,000 years ago, and in North China around 23,000-19,500 years ago**. These tools were used to process various grasses and roots. Tubers were important food resources for the Paleolithic hunter-gatherers and Paniceae grasses were exploited about 12,000 years before their domestication.

In the post-glacial took place a large expansion and migration of plants from southern tropical refuges (used during glacial) toward north, like from the Southeast Asia in the East and North China. Among grasses it was the wild rice (in many varieties) that turned into intensive exploitation by burning off the competing vegetation; this practice was in use for millennia without any attempt of domestication, and the adaptive syndrome of the annual wild rice has been greatly reinforced during such a long **proto-cultivation**.

Thousands of years before agriculture, the **seasonally mobile cultivators** of North China (on the middle reaches of Yellow River) and

Mongolia appeared (between 14,000 and 10,000 years ago, or even before). These first agriculturalists, being mobile, left very little or no archeological records. The first domesticated animals in East Asia, after the dog, were the pig (8,000 years ago) and boar (cattle).

People have first domesticated some animals as companions (dogs), and later other animals, which can be used for feeding. As long as they cannot feed the animals based on only local resources, they have to move them around, becoming pastoralists. The pastoralists would be encouraged to have temporary settlements in at least two locations, according with transhumant procedure.

The occurrence of some bad climate seasons (as a life experience) at the lower elevation place had compromised the natural habitat of plants, and forced the people there, from the selected healthy seeds of the remaining few uncompromised plants, to try diffuse them in the surrounding fields (leading to a life time commitment).

The eventual success of this enterprise had encouraged more trial and error processes, until the cultivation became stable and efficient. **This domestication type has been a necessity for survival** triggered by a limited time experience and eventually has been connected to some significant loss of the herding animals due to the same bad climate occurrence.

In this last instance the **settlement transformed from temporary to permanent, while this switch became justified by the plant cultivation**. The settlers were the hunter-gatherers or some nomad pastoralists. Because the plant cultivation has been seasonal, in the extra-seasons (non-agricultural seasons) the practice of herding and eventual hunting continued, or it has produced a separation among the family members, some being in charge with cultivation, others with herding. Most likely various strategies have been applied, while they were dependent on specific conditions of every locality. Some diffusion of practices have occurred, too.

Why a somehow a less-catastrophic change in climate caused the last resort strategy of plant domestication and cultivation that was more environmentally risky than the previous practice? The climate records did not indicate any catastrophe in climate in the time period 10,000-8,000 years ago, when the massive warming of the Holocene was under way.

Here something else happened: **it was a failure of the new linear thinking strategy** that misinterpreted a somehow limited climate event **because the nonlinear big picture was lost** and the very survival was left with one-single but linear option that was the manmade diffusion of seeds. This seemed to be the case because the plant domestication was random and very regional. Even not in all places the domestication of plants and animals occurred within the same time frame or like in the aforementioned correlations (in South America the animals were never domesticated by the local farmers).

Jared Diamond explained: "*Food production bestowed on farmers enormous demographic, technological, political and military advantages over neighboring hunter-gatherers. The history of past 13,000 years consists of tales of hunter-gatherers societies becoming driven out, infected, conquered or exterminated by farming societies in every area of the world suitable for farming.*" "**Food production arose independently in at most nine areas of the world** *(Fertile Crescent, China, Mesoamerica, Andes/Amazonia, eastern US, Sahel, tropical West Africa, Ethiopia and New Guinea). The homelands of agriculture were instead merely those regions to which the most numerous and most valuable domesticable wild plant and animal species were native. Only in those areas were incipient early farmers able to outcompete local hunters-gatherers. Once those locally available wild species had been domesticated and had spread outside the homeland, societies of homelands had no further advantage other than that of a head start, and they were eventually overtaken by societies of more fertile or climatically more favored areas outside the homelands.*"

Also Jared Diamond explained another important process; "*Expansion of crops, livestock, and even people and technologies tended to occur more rapidly along east-west axes than along north-south axes. The reason is obvious: locations at the same latitude share identical day-lengths and seasonalities, often share similar climates, habitats and diseases, and hence require less evolutionary change or adoption of domesticates, technologies and cultures than do locations at different latitudes.*" "*Food production was accompanied by a human population explosion...*

The sedentary lifestyle permitted shorter birth intervals. Nomadic hunter-gatherers had previously spaced out birth intervals at four years, or more. Food production also led to an explosion of technology."

Based on aforementioned information the farmers can have a population 50-66% larger than the hunter-gatherers. In terms of efficiency the hunter-gatherers can exploit some 0.1% of a given landscape, while the farmers can exploit 90% of it.

Jared Diamond: "*The expansion of agriculture that started from the original nine homelands had produced while expanding, from the same places, the genetic and linguistic elements. Domestication had been by far the most important cause of change in human genes frequency in the past 10,000 years.*" "*Skeletons from Greece and Turkey show that the average height of hunter-gatherers toward the end of the ice age was a generous 5'9" for men and 5'5" for women. With the adoption of agriculture, height crashed and by 3000 BC had reached a low 5'3" for men and 5' for women. By classical times (500-300 BC) heights were on the rise again, but modern Greeks and Turks have still not gained the average height of their distant ancestors.*" "*Life expectancy at birth in pre-agricultural community was about twenty-six years, but in post-agricultural community it was nineteen years. So these episodes of nutritional stress and infectious disease were seriously affecting their ability to survive.*"

However **farming encouraged and produced large inequalities, which manifested from mild (5,500 years ago) to already rampant (4,000 years ago) in less than 1,500 years**. It also produced a large inequality between sexes with the farmers' women as the most oppressed.

Population densities rose by 100 times within the farming areas.

Some of the other human groups from other distant or even near regions remained somehow pastoralists and/or hunter-gatherers for many thousands of years after the advent of agriculture in the Middle East (in Central Europe the pastoralist cohabitated with farmers until 5,000 years ago; the farmers came in this region 7,000 years ago). This can probably be explained by the nature of their land that was not very good for farming.

It was not an exclusivity in any of these practices: they were intermingled for a long time, while only at random the selection locally favored a process or another, or both. A study at the University of Hong Kong was focused on Chinese historical records on migration from 220 BC to 1910 AD. They found 928 nomadic migrations, from which 651 (or 72.8%) were directed southward, 243 migration northward and only 34 eastward-westward. The study found that southward nomadic migrations were due to changes in the amount of precipitations (humidity) in the local climate and very little affected by changes in temperature. The northward nomadic migrations were caused by the southern droughts.

At present the scientists speculate that the animals who come closer to humans, with each successive generation, gradually become a little more part of the household and human diet. Domestication phenotypes exist in a wide range of species, while it can be stimulated by natural selection (from among a collection of genes), conferring to the animal a natural type of tameness. The same genes exist in humans.

Hence we can say **that animals have habituated themselves to humans before any direct human interaction, implying a long period of unintentional management by humans**. Thus at the beginning of the domestication process only natural selection was at work; and only much later it was replaced by the direct and increased caregiving and by artificial selection.

There are debating dates on particular time of dog domestication. Most scientists agree on a date placed 15,000 years ago, while the place of occurrence is considered in the Middle East. However the rest of animals (15 species of large mammals from a total of 148) become domesticated from 10000 BC to 2500 BC (or according to other sources just starting with 7500 BCE). Many researchers suggest a paramount **role played by self-domestication, where the animals become gradually dependent on human help for feeding and refuge** (some animals preferred to stay close to humans, while other animals went away, producing natural separation between these two distinct behaviors).

However the domestication diffusion rate in Eurasia was 1 km/year that leave enough place for regionalism.

At its origin every ecosystem is a collectivity organized to conserve energy: it is naturally pre-programmed in this scope. The relationship between the ecosystem members is strictly designed for the scope of conservation of energy, where each member has an approximate natural quota for its own consumption.

Conjecturally, humans broke this natural rule when they were able to produce more energy than they immediately needed. Probably this first occurrence was connected with fire making. Since then the fire assured many other advantages: warming up, protection against other animals and/

or insects, cooking, tool making. Cooking, since 500,000-400,000 years ago, allowed steady increases in brain size and complexity.

The appearance of speaking gave to humans a far better advantage, allowing knowledge transmission and education. Every progress made by humans was translated in more energy acquired from the environment. At some point humans had an extra energy (in the form of food waste to be spear), assisting some animals for food and shelter, and gradually attracting particular animals to join human community. This process maybe started with scavenging of big game animals.

Before human intervention, the plants encountered a natural process of diffusion, where the agents, diffusing the spores, were the furs of animals, the birds, insects, the wind and running waters. **Nature uses diffusion as the transport vehicle that provides a balanced growth on a territory**. Pollination is maybe the best example: the bees transport the pollen between flowers and trees compensating for a job the wind cannot do it so well.

In cold climate the too much oxygen from the atmosphere can be diffused into the water, or if it is too much of oxygen in the water it will diffuse back into the air. The diffusion hence assures a needed balance of the environment, and where the elements, like plants and animals, can move in certain equilibrium within the ecosystem (like following a pre-planned dispersal along the architecture of the environment).

All biological systems employ fusion, as their main natural process, in order to acquire all needed substances for their own surviving. Fusion generates growth in biological systems: it combines various elements to produce the needed energy for life. Dispersal is only a natural vehicle for transport in the territory.

Climate changes suppressed or decreased some wild plant growth and productivity, endangering the life of hunter-gatherers. Instead of diffusing themselves (migrating) toward better resource-fitted environments, humans decided to apply the trick of seeds diffusion for growing locally the plants they needed; during a trial and error process, they succeeded the seed dispersal on their own. The process continued ever since.

In the above case the **manmade dispersal and selection were the linear forcing factors**, which since that time, diminished natural diversity of plants and animals. (For example today 12 species account for 80% of world food tonnage; the rest of 190 plants are not part of our main food chain).

10,000 years ago 200,000 wild plants were inhabiting the entire planet. Today the same number of plants occupy less than 40% of the planetary non-arctic dryland. This means "human zone" has exclusively dispersed or extended on over 60% of our planet's dryland. ***"Human zone" is not an ecosystem, but a "planar or linear *technology zone*", where we use machines to help do our work.***

Archeologists discovered the wild rice that grew in the current Yangtze Valley 13,000 years ago. The archeologists found that in the same valley the (domesticated) rice was grown in layers by 9,000 years ago. But they found traces of rice cultivation on the edges of the continental shelf dated 14,000 years ago. In the Middle East the domestication started in the Fertile Crescent in the era 8500-7500 BCE. As it seems, the first domestication in the Western hemisphere has occurred in East Anatolia and South Caucasus probably around 10,000 years ago.

From 200,000 wild plants only 200 have been domesticated before 2,500 BC and none after. **The entire process of domestication took less than 5,000 years.**

The domestication was a dispersal achieved in multiple stages and its origin was in human self-domestication that introduced the first kind of linear thinking; the linear thinking crisis produced the domestication of plants, while forcing into settlement; both independently occurred in various regions of our planet.

Why agriculture was a linear approach? Because it appeared as a result of manmade selection that ignored the complexity and natural rules of the ecosystems and environment. In the same time this was the first major approach to control the input-output linear mechanism. Since that time agriculture, economy and society at large try to control the same input-output relationship. The same linear logics appeared in building the permanent settlements. Both were placed among many other linearly driven mental addictions, which will follow further in the human development.

Before 4000 BC war was nonexistent; before 4000 BC, as the agriculture and urbanism developed, leading to some surplus for trade, the wars appeared as relatively rare and more ritualistic outcomes of growing conflict within the linearly causal agricultural society.

However warfare became possible when the societies turned structured, and when large agricultural productions were achieved, building up social and individual wealth. This revolution in **social structure that rested on new economic base of accumulated wealth was the most important factor responsible for the emergence of warfare**. As a mechanism of cultural development within agricultural-urban societies the **conduct of war became a legitimate social function, while war turned into a relatively normal event of social but linear existence**. The same input-output (or here occurred the gains versus losses games) concept was at work. It turned into the most widespread culture that continues all the way and unchanged into present.

13. Do we have "evolution"?

The reality of this change is far from simple and deserve comprehensive analysis.

There is no evolution; there is only environmental changes, which trigger adaptation. The selection coops with change only after the change took place: it reacts in response.

I will start with a consideration on the concept of "evolution". In their most honest thought the greatest scientists of 19th Century to 21st Century have shown that **time does not flow**, or that **it does not exist an arrow of time**. If this is the case than how can we analyze what we call the time flow? The only asset to employ at present into this analysis is the theory of **complex systems.**

The theory of complex systems indicate that very simple elements can generate, during an interaction, quite complex systems. A complexity evolves somehow on a vertical axis that I will call the "growth axis". On this axis we have hierarchies, where each hierarchy is characterized by its own scale and where each scale has its own physical laws (like subatomic, atomic, nanoscale and molecular). The stack of hierarchies generate the fractal, where each scale is self-similar to any other scale.

How evolves the complexity? It evolves by **self-organization**, meaning that the complexity is pre-programmed to adapt to changes in the environment or ambient. But the response of self-organization is always in the form of a plurality of options because all causes are complex by themselves; the response is complex. This indicates that each complexity evolves in direct and solely connection with the ambient, while it cannot be analyzed in linear manner because of its inner complexity.

The ambient is a complexity in itself, where are present elements like climate, cosmogenic, geomagnetic and telluric variables, but also vegetation and animal inhabitants. An entity immersed in this ambient, being it human or not, reacts to all changes by self-organization that generates adaptation.

14. Cosmogenic
Factor Revisited.

Recently I was reading an article published (2002) by Proceedings of National Academy of Science (PNAS) regarding a study on *"Natural radioactivity and human mitochondrial DNA mutations"*. This study was the joint venture of the scientists from University of Munster, University of Cambridge, University of Freiburg and University of Kiel.

The study was focused on a costal peninsula in Kerala (India) where **local community is subject to highest level of natural radioactivity in the world.** The article says: *"The increased radioactivity is due to the local abundance of monazite, a mineral containing cca. 10% thorium phosphate. The radioactive strip measures an area of 10 km by 1km, but supports a population of several thousand. The **biologically effective radiation dose received by costal population is ... 10 times greater than the worldwide average**". "Studies in Kerala did not reveal significant abnormalities in local residents".*

Here we can compare with the situation in the Arctic area (described in previous chapter on cosmogenic influence), where the average natural radiation is 50 to 100 times higher than average in temperate climate, and no significant health problems were detected. In fact, local health in the Arctic was much better than in temperate climate areas.

However the aforementioned study concluded: *"the **radiation-associated mutations have hit nucleotide positions that mutate, on average, 12 times faster during evolution**. The observation that **radiation accelerates point mutations** at all is unexpected, at first glance, because radiation was, until recently, thought to generate primarily DNA lesions. ... there radiation-associated point **mutations are also evolutionary hot***

***spots**, indicating that the radiation indirectly increases the cell's normal (evolutionary) mutation mechanism".* "As demonstrated, **our mtDNA results strongly support an acceleration of the evolutionary DNA mutation mechanism through radiation".*

We can assume that a similar radiative effect occurs in the Arctic, causing accelerated point mutations.

This article finally brought to the public the scientific proof that **some of the cosmogenic radiation causes mtDNA evolutionary mutations,** and not cancers and any lethal or serious diseases occurred. The same study indicated that a low radiation zone (located in the same area) produced only one evolutionary mutation in mtDNA, while the **strong radiation zone produced several mutations** of the same type.

Another article by P. Andrew Karam, at University of Rochester (US), analyzes the situation at Ramsar, in northern Iran (on the shore of the Caspian Sea). He explains: *"The high background radiation in the hot areas of Ramsar is primarily due to the presence of very high amounts of Ra decay products, which were brought to the earth's surface by hot springs. Radium is dissolved from the rocks by hot ground water. The **radioactivity in local soils and the food grown in them is also high**. There are 9 known hot springs with various concentrations of radioactivity around Ramsar. **Residents and visitors use these springs as health spas**. Iranian researchers have calculated maximum credible annual radiation exposure of the area up to 260 mGy, while the recommended dose limit for workers in Iran is 20 mSv per year. The people who live in high radiation areas of the world are of considerable interest because they and their ancestors have been exposed to abnormally high radiation levels over many generations."*

He explains the results of the scientific investigation: *"**Preliminary studies (2002) show no significant differences between residents in high background radiation area (HBRAs) compared to those in normal background radiation area** (NBRAs) in the areas of life span, cancer incidence, and several other issues. Specifically HBRA inhabitants had 44% fewer induced chromosomal abnormalities compared to NBRA residents." "Similarly data obtained from **studies on HBRA inhabitants of Yangjiang, China, and Kerala, India, show no harmful impact induced by regional natural radiation**. These results suggest that exposure* *to elevated levels of natural radiation does not result in increased chromosomal damage and is not detrimental to the health of residents".*

In Guarapari (Brazil, where it is a popular touristic beach) the annual exposure to radioactive radiation is 175 mSvs and in Ramsar (on Caspian Sea shores) is 250 mSvs (**80 times higher than the world average**, hence similar to the Arctic). In **Yangjiang** (next to the South China Sea) the residents live with an **annual exposure three times higher than the world average**. In all four places, and also in the aforementioned place from Australia, **people consider these hot spring resorts to have healing power**. In all these places the cancer occurrence is lower than on average. A study revealed that local residents exhibit significantly **higher expression of the CD69 gene responsible for the production of lymphocytes and natural killer (NK) cells**.

So we have a collection of three studies (on Kerala, on Ramsar and on Yangjiang), plus the situation in the Arctic, which make a **good case for a favorable influence of the cosmogenic radiation on human DNA and our brain mechanism** (the brain rhythms).

Returning to mutations, the quantum superposition of gene mutation (within quantum theory) **do not manifest immediately in the phenotype of organism, but accumulate in gene pool of species during long periods of time. Natural selection selects from the pool those genes that are available to better fit changes in the environment.** So, the high natural radiation zones seem to act on our gene pool, sometimes adding new genes, which are being used only when needed, and sometimes only selecting those existing genes capable to fit the best a particular forcing (like selecting more primitive adaptations, which may have a better endurance).

Also recently it surfaced that **low energy amounts of galactic radiation interferes with radiation emitted by the DNA (UV radiation) that is produced by the DNA weak laser**. The geomagnetic shield and the **atmosphere play the role of an interacting filter (a conversion role) that degrades (absorb) light rays and strong radiation, transforming them into ultrasound phenomena.**

Three particular elements will influence this cosmogenic radiation penetrating to the ground. First is the phenomenon of **geomagnetic excursions** that diminishes the geomagnetic intensity and shielding over the high latitude areas (above 45th parallels north and south). Such excursions are quite numerous and are connected with climatic and volcanic episodes and also with changes in vegetation cover and animal inhabitation.

The second element that favors the radiation to reach the ground is the geomagnetic cut-off produced by the main **geomagnetic anomalies** of East Siberia and South Atlantic. At present we can estimate that these two anomalies have considerably migrated over time. It is estimated that the South Atlantic anomaly was above South Africa around 70,000 years ago. Today it is above the southeastern coast of Brazil. Several decades ago it was over the South Atlantic.

The East Siberian anomaly, during an era before the LGM (around 40,000-30,000 years ago), was estimated to exists above the Yellow Sea/Korea/Japan location. It was above the Yenisei Basin/Altai Mountains from 20,000 to 25,000 years ago. Around 26,000 years ago (before LGM) it was above Lake Baikal. At present it is east of Lake Baikal.

Again, this cut-off of shielding allows the radiation to reach the ground in locations like the aforementioned ones. **Whatever is the type of radiation**, reaching the ground (being light, sound, etc.), **and its own frequency can interfere with our DNA frequency by resonance**. The result can be an annihilation or an augmentation of the original DNA's features. This can spell mutation.

The third element is the **configuration of the geomagnetic field that is influenced by the geomagnetic jerks**. Also this variation in geomagnetic core configuration may produce **telluric radiation** variability with significant local influence on the ground.

The aforementioned geomagnetic variability has created a plenty of cosmogenic interferences on the ground of our planet. These interferences were significant for animal DNA and also for the vegetal cover, eventually contributing to mutation as a form of adaptation.

Russian scientists have determined that a **cosmogenic radiation penetrating through a cut-off can successfully produce a mutation in an individual if the exposure is at least 55 years long, and if the individual is able for procreation after this exposure**. This means that here is a barrier of age: the human exposed to such cosmogenic radiation since his birth must live at least 55 years of age and make children at this age. Then the children will have this mutation. Is this the aforementioned case from Kerala, where people live there for hundreds and thousands of years? Or is this the case of the Paleolithic people living for thousands of years in the Siberian Arctic?

A study (2011) of Valery and Tatyana Glazko, from Russian State Agrarian University, was investigation the results of the **Chernobyl nuclear accident** from 1986. The study was performed on animals and humans from this specific zone. It was found that only 5%, of 116,000 people evacuated from the zone, absorbed more than 100 mSv per years (a dose three times lower than in Ramsar, Iran). They found that "*the real danger does not arises from the absorbed radiation dose itself but rather from its "novelty" for the exposed population, species, or community*". "*… there was no significant increase in the number of constitutive mutations carriers in the studied populations. Therefore, **the main response to chronic low dose radiation is not in induction of new genes appearance, but rather in the preferred selection of genes (activation of the non-active genes) combinations accumulated during the previous generations. A change in selection conditions leads to preferred reproduction of the least specialized forms (which are more primitive forms). The result of Heiervang et al. that in utero irradiation among young people led to verbal IQ decrease** could also viewed from this perspective, as the verbal IQ is an evolutionary later acquisition compared to nonverbal IQ based on ancient structures*". "*… the increased level of ionizing radiation just raises the frequency of occurrence of cytogenetic anomalies*", which did not pose a health threat. It was found out that the irradiation is "*followed by **speeding up of genetically defective cells elimination***". The result in mice "*was an **accumulation of radioresistant individuals** about 26 generations after the accident*".

Another study, based on Hiroshima and Nagasaki (Japan) nuclear bombing case, found that the radiation from August 1945 increased the number of the genetic effects in the offspring (humans and animals). The radiation did not cause any unique effects because the seen enhanced effects (in the offspring) already existed in a latent form in all unexposed individuals or animals. This confirmed the idea and the results, showing the **effect of the preferred selection that occurs within an existing pool of genes.**

In my opinion the Arctic radiative background and other radiative conditions existing for the previous Paleolithic population of Siberia favored a suppression in the verbal-linear-content-thinking, stimulating the older abilities of the visual-context thinking. In the course of time, ultimately this feature had driven toward the option of developing the **Indo-European** family of languages (as the synthetic group of the **fusional languages**), as opposed to the agglutinative family of languages developed in the Middle East. There is interesting to observe that the Mongoloid languages (as the analytic group of languages) are closer to the Indo-

European languages, but far distant from the agglutinative languages. This is so because the fusional languages are in permanent process of transition toward analytical, while the analytical also has a recent trend toward fusional approach and then back. The agglutinative languages tend also toward analytic and fusional, and then back.

Chinese (as a Sino-Tibetan) is an analytical language that combine contextual and analytic qualities. Learning Chinese is about learning to understand its contextual fundament, or analyzing this complex context (finding a relevant meaning out of the context of possible meanings). The context describes comparative patterns of meaning. In other words, the Chinese language originated from a nonlinear perspective. From the same perspective originated the fusion languages (Indo-European). The agglutinative types are the result of centrifugal tendency in thinking and constructing.

Another study came to the conclusion that radiation hormosis model shows that **chronic low level of ionizing radiation, in addition to background radiation** (that is the case in the Arctic), **are beneficial by activating repair mechanism that protect against disease**.

The aforementioned barrier of age implied that "moderns" or Neanderthals, subject to cosmogenic mutation, must live some 60 years of age and have children after the age of 55. With current archeological information about the age of skeletons it is hard to determine when this moment occurred in our pre-history. As a simple speculation, it seems that this barrier was successfully passed by homo some 41,000 years ago, during the Laschamp excursion.

The duration of the most geomagnetic excursions is longer than several thousand years. **During last two Ice Ages (over 210,000 years**

long) we had 14 major excursions, and geomagnetic field was in excursional mode some 25% of the time. So chances for cosmogenic radiation influence on humans living in places above 45 degrees latitude is a scientific fact.

The irregularity of the cut-offs over Earth have created local patterns, which may have been responsible for a type of local climate and climate refugees.

All aforementioned variations have provided the basic elements, which have **produced complexities in an irregular manner.** Each complexity has evolved on its own, while the end-result has been a graduation to another hierarchy. **Once a complexity reached new hierarchy, its aspect changed, while it remained self-similar.**

Maybe for the human observer the change of hierarchy is interpreted as a dissolution and disappearance of certain type of organization. Self-similarity may appear to someone, anyone only as a large distinction, differentiation or even like a completely new product because self-similarity is evident only when the process is completed, and when one benefits from a proper vantage point.

Can self-similarity be shown on both vertical and horizontal scales? Probably this is the case, and then **we see progressing evolution where in fact is only self-similarity in all directions.** If this applies to human "evolution", it means that it has been not evolution but only self-similarity because the basic elements, which created each complexity, were self-similar, too. We may have an overlap of self-similar conditions, where the **differentiation between final products is seen by linear thinking as a gradual evolution and natural selection.**

In my mind, there is no question that our interpretation of history is far from complex but natural reality because we ignore the context of many powerful alternative views, and then we cannot select from this context which view is more appropriate to this complex reality. Everything is connected to changes in the functioning of the brain, which mostly occurred in the last 15,000-10,000 years, but more marked in less than 8,000 years. Our current industrial revolution started less than 500 years ago. Our computing assisted revolution started 60 years ago. **We have 8,000 years of progressive linearity, just opposing the complexity of reality, and nothing more.** In the last 40 years we discovered the idea of the complex systems, but we still do not know exactly what they are. We are another complex system and still do not know about it.

Today we have the tendency toward verbal processing that brings us linearity in thinking. We had lost, or we had mentally suppressed, a large part of the previous visual thinking abilities, which fact drive us toward "**verbal primitivism**". Can someone compare this to current "**all time texting**"?

15. THE ROLE OF DIET.

In 2010 a paper by K. Kochman and M. Czauderna with Polish Academy of Science showed: *"Both omega-3 and omega-6 fatty acids are crucial elements in the structure and function of cellular membranes, determining their proper physiological activity in regard to fluidity, intracellular transport, and protection against intruders such as bacteria and viruses. These acids actively participate in the biosynthesis of such neurotransmitters as dopamine and serotonin, which are required in nerve cells for quick and efficient signal conductance. The proper content of omega-3 fatty acids in diets increases and improves learning ability, problem solving skills, concentration, memory and communication between nerve cells. Omega-3 acids support positive mood and emotional balance, good mental skills in aging people".*

The same study says: *"Scientific studies in the 1970s on the health of* **Inuit-Eskimos** *living in Greenland clearly showed that they were not only the healthiest population in the world, but* **their good health exceeded by far that of any other population in the world***. Their diet consisted mainly of whale, salmon and seal. The incidence of such serious diseases as coronary heart disease, arthritis, diabetes and many other was extraordinary lower than in other populations.* **They also had the lowest infant mortality rates in the world.***"*

Sources of these fatty acids include*:* salmon, mackerel, vegetables, vegetable oils, linseed. DHA role is paramount for the development and function of human brain *(*its mass being more than 50% fat, and contains an abundance of omega-3 fatty acids*).* ***"These fatty acids activate syntaxin 3, allowing growth of neuronal extensions that is an essential process in brain development."*** *"Fatty acids imbalance* (like fatty acids deficiencies*) can adversely affect brain development, including the ability to respond to environmental stimuli."*

Their study also showed that a similar situation of very good health have arisen in the past in Iceland. We found other studies that indicate similar results for Siberian populations.

The Siberian environment with subarctic and arctic climate has several advantages: **the cold air has more oxygen than warm air, hence it is better for human organism. The cold waters in cold air area have more oxygen, which assures minimal micro-organisms into it, gives a better taste, and support a large fish population. Human bathing in cold water strengthen the health**.

Many other **plants from Siberia are famous for their therapeutic and nutritive qualities**, like Siberian Pine Nut Oil, Chaga (Inonotus Obliquus) wild mushroom, Siberian Ginseng, many types of local berries, various tree leaves and grasses, etc. The Science Academy of Siberia **researched this local diet** for possible neurological effects; it **was found to cause significant and favorable impact on intracellular connectivity and synaptic firing range and frequency. But most important was the finding about major assistance in the brain's rhythms**.

Stuart Wolpert describes the findings of a study (July 2008) led by Gmez-Pinilla at UCLA saying that Omega-3 fatty acids were found to **support synaptic plasticity and positively support the function of learning and memory**. He also shown that the research confirmed the hypothesis that **health and mental health can be passed down through many generations**. *"Evidence indicates that what you eat can affect your grandchildren's brain molecules and synapses"* (Gmez-Pinilla saying). *"In contrast to the health effects of diets that are reach in Omega-3 fatty acids, **diets high in trans fats and saturated fats adversely affect cognition**." "Folate supplementation, either by itself or in conjunction with other B vitamins, prevent cognitive decline"* because *"adequate levels of folic acid are essential for brain function." "Smaller food portions with the appropriate nutrients seem to be beneficial for the brain molecules, such as BDNF."* Here **animal foods are the only source of B 12 vitamin**. It is shown that vitamin B 12 deficiency has severe neurological consequences.

These aforementioned studies are the best indication we can find **to illustrate the situation of people living in Siberia 26,000 years ago**. I refer to Mal'ta boy of Lake Baikal, who's on the record with the first gene mutation to haplogroup R. Another Siberian individual from Afontova

Gora (Krasnoyarsk) had the first R1 gene around 20,900 years ago. Their relatives, settled around 15,000 years ago somewhere on the north shore of the Caspian Sea, have produced the splitting of R1 into R1a and R1b, which both subclades form today the majority of the population of Europe and the ex-Soviet Union, but also are present in many other countries of the Eurasia.

These **Siberians living 26,000-20,000 years ago were hunters and fishers and had the same diet that of the Inuit people described in Greenland in the 1970s.**

However the article did not mention that the health of today Inuits has been badly affected by new diet with carbohydrates and other modern practices.

The successors of the haplogroup R, who were the R1a, or the Aryans, returned 20,000 years later to their ancestral lands from Siberia and formed the Afanasevo culture of Minusinsk basin, Krasnoiarsk and Altai Mountains regions of West and Central Siberia (3,500-2,500 BC).

During this long period of time, their movement was from Central to West Siberia, to Urals and Caspian-Pontic regions, Caucasus region, entering Iran and Anatolia, from where they crossed into the Balkans. In the Balkans originated the new Indo-European language. The Aryans, speaking the Indo-European language, created series of cultures on their way back from Balkans to Siberia, like the proto-Hellenic (6,500-6,000 BC), the Starcevo-Vinca-Cris (6,200-5,000 BC), Cucuteni-Trypillian (5,000-3,000 BC), Khvalynsk (5,000-4,500 BC), Sredny Stog (5,000-4,000 BC), and Yamna (3,600-2,300 BC). From Afanasevo they formed some other cultures like Catacomb culture (2,800-2,200 BC), Sintasha (2,100-1,800 BC), Andronovo (2,000-900 BC) and BMAC. And then they conquered (1,700-1,650 BC) Iran and Indus Valley civilization around 1, 550 BC.

According to the genetic findings, a branch of Aryans (at that time they did not speak the Indo-European languages, which were form later in the Balkans) entered Iran and India for the first time around 6,000 BC, while another branch entered almost simultaneously the Eastern Balkans (Greece). The branch that entered India advanced to Center and South India, but also crossed to the east until reached the Mekong Valley in the Southeast Asia; they left few genetic traces from Laos to Thailand and to

Cambodia. Some specialists call the first Aryan invasion of India as the original Vedic Aryans. The issue is very disputable because few Aryan genes were also found in the Nepalese and Tibetans, and also in North China, Manchuria, and Korea.

The most interesting question here is **why the Aryans returned to their ancestral land? The answer may be in their ancestral memory about the** *"healing grounds of Siberia"*, **which grounds heal by the locally imposed diet, but also** *this diet empowered the body and the mind.*

However the Siberian bearers of the haplogroup R1a and R1b invaded and conquered most of the farming settlements of the Eurasia between 3,000 BC and 1,500 BC, imposing their genes into higher segments of the local population (like in the aforementioned discussion on India). The R1a were called the Aryans, and they came with "space (visual) intelligence", nonlinear thinking, domesticated horses, chariots and metallurgy, helping the local farmers to upgrade their "civilization". But later, **their original communitarian faith changed being influenced by the habits of the agricultural society; their pastoralist lifestyle habits were imposed to the agriculturalists by forcing them to introduce the patriarchate; from that point on, everything changed in the agrarian society, opening the way toward aristocratic elites, despotism and social inequality.** This was the consequence (as the adaptation of the Aryan elites to local agrarian behaviors) of changing their pastoralist but visual, contextual, nonlinear thinking into the verbal, content-analytical, linear thinking of the agrarian societies they conquered. In the meantime the R1a always travelled with smaller groups of R1b (named Arbins), while suffering their interference.

However among the Aryan groups, those producing Andronovo culture, remained approximately (some moved west from the Ural River valley to Volga River valley, eventually settling in Donbass) in the same region for a long period of time (over 1,200 years), while preserving the original pastoralist mentality and lifestyle. This branch of Aryans seem to be the core from which later evolved the future proto-Balto-Slavic people and the proto-Slavs.

Here I like to quote a more authoritarian opinion from Eupedia on the R1a haplogroup whereabouts: *"R1a haplogroup…have been the dominant haplogroup among the northern and eastern Proto-Indo-European language speakers that evolved into the Indo-Iranian, Thracian, Baltic, and Slavic branches. The Proto-Indo-Europeans originated in the Yamna culture (3300-2500 BCE). The southern Steppe culture is believed to have carried predominantly R1b lineage, while the northern forest-steppe culture would have been essentially R1a dominant.* **The first expansion of the forest-steppe people occurred with the Corded Ware Culture.** *The migration of the R1b people to central and Western Europe left a vacuum for R1a people in the southern steppe around the time of the Catacomb Culture (2800-2200 BCE).* **The forest-steppe origin of this culture is from the Corded Ware Culture.** *The* **Balto-Slavic** *and Indo-Iranian* **language groups** *belong to the same Statem isogloss and both appear to have* **evolved from the Catacomb Culture** *(Ukraine). Ancient DNA testing has confirmed the* **presence of haplogroup R1a1a in samples from Corded Ware culture** *in Germany (2800 BCE), from Tocharian mummies (2000 BCE) in Northwest China,* **from Kurgan burials (1600 BCE), from Andronovo Culture in Southern Russia and Southern Siberia, as well as from a variety of Iron-age sites from Russia, Siberia, Mongolia and Central Asia."**

Nevertheless **Siberia continued to produce somehow large groups of population for several thousands of years, who were more intelligent and free of infantile mortality because of their diet; it was a continuity of waves after waves of smart and healthy people, who, migrated westward and invaded Europe;** in the course of time they have been gradually absorbed into the modern European population.

16. The case of exposed continental shelves of East Asia

On the opposite side of the Asian continent, into its southeast, a mutation created first Mongoloid 30,000 years ago. However this type of the Mongoloid did not become similar to modern counterparts until 10,000 years ago.

The East and Southeast Asia, during the LIA, have large continental shelves, which become exondated, turning into a good dry land for human use. Even during the harshness of the LGM the coastline of these new lands were covered with coniferous forests from Taiwan to Korea and to Japan. In many places, toward the interior, some prairie landscape and forest-steppe appeared. In some respects the landscape was similar to the Siberian one.

Here I will quote two recent studies. The first study is done by a Chinese team (30,000-year vegetation and climate change around East China Sea shelf inferred from a high resolution pollen record) in 2010, and it shown: *A high resolution pollen record derived from core DG9603 reveals vegetation and climate changes on the East China Sea Shelf (ECSS) in the past 30,000 years. From 29.8 to 26.6 cal ka BP the ECS was covered by warm temperate forest-steppe and wetland. During the period 26.6-14.8 cal ka BP wetland and temperate forest-steppe developed around the ECSS. From 14.8 to 5.3 cal ka BP sea-level continuously rose, and ECSS was gradually submerged. In some exposed areas of the ECSS and lower reaches of the Yangtze River, northern subtropical forest developed instead of temperate forest-steppe and wetland. The pollen record shows that the rainfall and temperature increased continually during the period 14.8 to 12.8 cal ka BP, reaching the level of Holocene optimum."*

In the second study (2013), besides the Chinese researchers, it were Austrian and Japanese researchers. It concluded: *"The area of study comprise the warm-temperate deciduous forest (WTD), as presently found scattered in mid-elevation subtropical (Central/South/East) China, predominant in low-elevation North China and Korean Peninsula, and disjunctively distributed in main islands of Japan. Fossil pollen analyses have indicated that during the LGM (21-18 ka.) the habitat of East Asian WTD forests in the northern part of their range (North China, North Japan) contracted. However the paleo-biome reconstructions suggest that these forests also expanded across the large expanses of continental shelf (c. 1 million square kilometers). It is now widely believed that a near continuous belt of WTD forests spanned the ECSS continental shelf during the LGM (and possibly in earlier cold periods), connecting the presently disjoint populations in East China, South Japan and the Korean Peninsula)."*

These two studies indicate that **the entire East China Sea Shelf was covered by a temperate forest and forest-steppe that was connected with currently existing WTD forests from China, Korea and Japan. In other words, this huge forest covered some 2 million square kilometers for almost the entire duration of the Last Ice Age (LIA) that was 100,000 years long.** Probably this forest and its associated forest-steppe were connected through Korea, Manchuria, Primorie, East Mongolia and the Amur Basin with the **South Siberian temperate and humid forest (taiga).**

Because of the attractiveness of this coniferous forests the Mongoloid inhabitants from subtropical South China moved to the coast of ECSS (probably first around Hainan), settling at the mouth and on the estuary of local big rivers (Huang He, Yangtze, Pearl River, Hai River, and so on). Probably they were busy fishing, and hunting, for the entire duration of LGM.

Probably they moved to ECSS (during the optimum climate of LIA) shortly after their arrival in South China (around 38,000-36,000 years ago). The study of the skeletons from Okinawa indicated that the population living there from 35,000 to 28,000 years ago had a dentition similar to people from South China and opposed to those living in Japan and Korea. This fact seems to suggest that the people from Okinawa, Ryukyu and Kyushu migrated there from the ECSS.

Their diet was in part similar to that of the Siberians (excepting the cereals). Being on the coast, they probably explored the area to the north and south, reaching into Japan and Beringia, and to the south and east to the Indonesian-Filipino Archipelagos, which both at the time had significant land bridges and large exposed shelves.

Their venture into north have led to the Arctic zone colonization sometimes 18,000 years ago. Also they crossed Beringia and entered Alaska, colonizing Americas (13,000-12,000 years ago).

The mental skills of the Mongoloids were shown at an early time. For example there is the engraved stone from Ningxia (middle reaches of Huang He) dated back to 30,000 years ago, showing the use of symbols. In South China a pottery 20,000-18,000 years old was discovered not far from a place where the 2,200 years old the best conserved mummy was found. The tortoise shell engraving in East China from 8,600 years ago was the first writing. The first compass was produced 4,600 years ago in Central China.

A Chinese study (2002) on an archeological discovery (1989-1991) of early rice remains in the middle-lower Yangtze River and upper Huaihe River dated these domesticated specimens at 8,000-9,000 years old. The study said: *"By counting the proportions of wild and domesticated rice phytolith fossils from the Diaotonghuan cave in northern Jiangxi Province along the middle Yangtze River, Zhao (1998) identified **a rice horizon dated to 13,000 years ago**. With this age constraint, the rice would be the **earliest domesticated cereal crop in the world**."*

The study analyzed the climate changes in East China Sea (having a surface of 770,000 square kilometers) in the last 20,000 years, which influenced the habitat conditions on the exposed continental shelf of East China Sea. The results were the interpretation of deep sea cores drilled in this shelf. The core samples were examined for evidence of phytoliths of domesticated rice. **The fossil record revealed a sudden appearance of phytolith rice at about 13,900 cal. yr. before present that corresponds in time dating to Zhao discovery from 1998 on the aforementioned current mainland, in a cave.**

The rest of the study analysis referred to several other climate aspects and other organic samples from the same core (DG9603 core from 1996). **The core was collected at the paleo-eastern-limits of the exposed continental shelf; it was collected from the paleo-estuary of Yangtze River placed 600 km east of current mouth.**

The study concludes: "*The rice phytoliths occurred first in this area at about 13,900 cal. yr. BP, or during the time span* **13,900-13,000 cal. yr. BP; warmer and more humid conditions probably favored rice domestication. During the period 13,000-10,000 cal. yr. BP the climate was cooler and drier than today, which might have a great influence on human activities and explain the disappearance of rice phytoliths at this time. After 10,000 cal. yr. BP the climate conditions became humid and warm, and the early rice domesticators became true farmers.**"

This study and its information are important from several points of view:

-first it indicates that rice domestication around 14,000 ca. yr. BP was wide spread from current mainland all over the exposed continental shelf to its coast of that era that was 600 km east of the Yangtze current mouth (the sea elevation was -100 to -90 m below current elevation); it may not was the entire shelf but it certainly was along the paleo-Yangtze Valley; 14,000 years ago still the entire shelf was exposed;

-the Younger Dryas cold episode may interrupted the rice cultivation, but it still allowed the cultivation of other types of plants (cereals like barley and other some more cold resistant plants); eventually the Younger Dryas forced the farmers to migrate south along the shore (toward the Pearl River mouth); because of the cold episode the progress of the sea flooding, caused by deglaciation, ceased or stopped in most places, while in few places it was recorded a sea retreat (the process took at least 1,000 years);

- around 10,000 years ago the sea already flooded some 40% of the shelf that existed as exposed 15,000 years ago; at this moment in time the Yellow Sea was formed only like a large gulf elongated on the south-north direction; this new branch of sea basically flooded a previously very large and very long estuary (that contained the flow of Huang He and many other rivers);

-secondly it shows that **a considerable population of farmers existed on the exposed shelf since 10,000 years ago**, while the domesticated rice cultures continued after the Younger Dryas (probably a 1,500 to 2,000 years interruption);

-thirdly it suggests that, **during the Late Paleolithic and the entire Mesolithic until early Neolithic, the Chinese population was densely living on the exposed continental shelf**; such dates of early inhabitation of "moderns" do not exist for mainland (except the aforementioned cave). However this explains that **during the gradual sea water flooding this population from the sea-flooded-shelf moved to current inland**.

The sea level still was -60m around 10,000 years ago and +2m 5,500 years ago (3500 BC); so **the migration out of the shelf ended 6,000 years ago (4000 BC)**. All this population of farmers first became more densely packed on the shelf, but later, or 8,000-7,000 years ago, when the density turned too much to bear, and the exposed shelf was mostly consumed by the sea waters, the people from the shelf begun to reach the mainland in large numbers.

Another aspect is of major importance: it is connected to **Kyushu and Ryukyu volcanos** from Japan. For example the Aira Caldera (near Kogoshima in Kyushu) has a diameter of 20 km and was created by a huge eruption 28,000 years ago. This eruption (magnitude M 8.4) produced volcanic ash and pyroclastic flows, which covered extensive areas. Human inhabitation started in Japan 34,000 years ago only in the Kyushu and Ryukyu, and was severely disturbed by the aforementioned eruption. Another large eruption (Kikai in the same caldera) took place 6,350 years ago (4350 BC).

Both eruptions have been responsible for alleged **tsunamis** because a tsunami is formed when there is ground uplift and quickly follows a drop. The tsunami can be caused by caldera collapses, tectonic movement from volcanic activity, and pyroclastic flow discharge into the sea. As the wave is formed, it gains great speeds in deeper water (1.000 km/hr.) and still speeds in the shallow waters (over 320 km/hr.). Such tsunamis, along several others produced by the Okinawa volcanos, would be of devastating proportions for the on-shore and riverine inhabitants placed on the opposite shore of China and Korea.

We need here a small discussion. During the entire Last Ice Age because of ice accumulation, the global ocean elevation dropped significantly: most of this time the drop was in the range of -50 m to -135 m. This implied that **all continental shelves of this planet were exposed (dry land) for most of this 100,000 years long Ice Age**. The interglacial eras are each only 10,000 years long. Hence in the last 2 million years only in some 200,000 years the Earth had current look (that is 10% of the time). So **Glacial was the rule, the interglacial was the exception**.

From all continental shelves, the East China and Yellow seas have the widest shelves (600 km). The sum total surface of these seas is over 1,600,000 square kilometers, and 80-90% of this surface is continental shelf. So **during the Last Ice Age at least 1 million square kilometers of Chinese continental shelves were exposed during the cold episodes, and more than half of this surface was exposed most of the time (650,000 square kilometers)**.

Different studies on these shelves indicate that **during the LGM grasslands and arboreal landscapes were developed inland and along the coastal area of the shelf**. The coastline until 17,000 years ago was along a line that united northern tip of Taiwan with Cheju Do Island and Korea mainland. Kyushu-Shikoku-Honshu formed one single island, and Sakhalin-Hokkaido formed a peninsula out of Primorie (Russia Far East).

The continental shelf's coastline was separated from Ryukyu Archipelago by a 200 km wide sea (Okinawa Trough) that extended 900 km from south (Taiwan) to north (Kyushu Island). The archipelago had also a lot of exposed land. **South west of Okinawa Island was discovered a coral reef development that begun around the LGM and shows that the climate was warm enough to support reef growth and evolution**.

There is interesting to know that the **first "moderns" in Japan were found only in Okinawa and existed only in the era 34,000-28,000 years ago** (an interruption in inhabitation occurs from the 28,000 year old volcanic eruption to 26,700 years ago, when the inhabitation resumed).

Their eventual successors produced the 12,000 year old petroglyphs and 15,500 years old pottery. Who were these people and from where they came from? Definitely they crossed the Okinawa Trough, coming from the exposed continental shelf of the East China Sea.

Next wave of "moderns" belongs to Ainu and Jomon cultures (16,000 years ago), which populated originally central-south regions of Japan, but later moved north. They created one of the oldest pottery 16,000 years ago (similar with that found in Primorie). South China pottery was even older, being dated as 18,000 years old.

Let's return to our subject here: the shelf of East China Sea. The lowest sea level position was 22,000 years ago (LGM) and **post glacial marine transgression started 19,000 years ago**. We have now information about the first domestication of rice which occurred 13,900 years ago at the paleo mouth of Yangtze River. This paleo mouth was on the shelf somewhere 600 km east of current mouth; the research on another Chinese river, on Huang He, shows that its mouth was 11,000 years ago into a sea with the elevation -80 m. At 10,000 years ago the mouth of Huang He was at a sea with elevation -60 m. So the transgression was 20 m in 1,000 years.

In the Fertile Crescent the domestication of plants started 12,000-11,000 years ago. Thus the rice cultivation on the exposed shelf of China (13,900 years ago) will be the earliest domestication of cereals that we know of (it was ahead of the Middle East with 3,000-4,000 years). However, **theoretically we should have here, in East Asia, some forms of large settlements, like in the Middle East (Jericho) or in Balkans (Lepenski Vir)**.

In the meantime the dispersal of this early culture into current mainland have not occurred, or the evidence of this is not available, yet. In south China was discovered a 20,000 years pottery and in Ningxia was discovered an engraved rock dated 30,000 years old. Maybe this is the beginning of finding the dispersed elements of an early culture developed on the shelf. **Taiwan Island was part of the shelf** (until the sea level was shallower than **-80 m**) **until 11,000 years ago**, so there must be hiding many clues about what was developed under the current waters of Yellow and East China seas and eventually dispersed into Taiwan before 11,000 years ago.

It is estimated that the population of humans on our planet was 4-6 million people 10,000 years ago. How many of these people lived on the world wide exposed continental shelves? 4,000 years later (6,000 years ago) the world population risen to 60-70 millions. This was called the Neolithic explosion.

At present the alluviums (sediments) are so thick that no archeological work is possible on the shelf. Only the drilling into the sea floor can bring some very limited archeological information and eventual evidence.

My point here is that **China may justifiably claim to theoretically-have the first cities** (while they would remain unknown for a very long time or forever) **of the world and a considerable pre-historic but social development on its continental shelf.** There is to assume that this culture on the exposed shelf has had crossed the Okinawa Trough and settled in the Ryukyu Archipelago. This event was triggered by the **climate optimum moment of the Last Ice Age (38,000-35,000 years ago)**, while it was not possible before because of a cooling event 40,000-39,000 years ago (generated by the Laschamp excursion from 41,000 years ago).

During LGM it was created like a barrier between Eastern and Western parts of China. As it appeared, **the shelf and East China were the most favored by the "moderns" who probably came to South China around 50,000 years ago and who became at that time an "Asian type (40,000 years ago), while they separated from the Cro-Magnons (European) type somewhere in the Middle East.**

To conclude I would say that an early human culture on the shelf was subject to frequent sea and river flooding, while the large volcanic eruptions from Kyushu might produce devastating tsunamis and other effects. In Kyushu are 9 active volcanos and are documented massive eruptions, 8,000, 6,500, 4,850 and 4,600 years ago.

Maybe large numbers of people living on the exposed shelf perished due to sudden environmental, climate and geophysical events (earthquakes) before they were able to **move on the mainland around 6,000 years ago.** Here my argument comes from the fact that all Neolithic cultures of China (except Nanzhuangton, placed 8,500-7,700 BC) started to appear at sudden around 9,000-7,000 years ago (or 7,000BC-5,000BC). This is the moment when the exposed shelf diminished dramatically.

Another fact is connected to the area where these cultures spread: they were mostly dispersed along Huang He Valley, secondly in lower Yangtze Basin and thirdly in the area of the Pearl River mouth. **By far the Huang He basin colonists were the most in numbers and areal extension.**

From 9,000 years ago to 7,000 years ago Huang He river moved its lower course from Bohai to south of Qingdao, and just north of Shanghai. When the sea reached current elevation, the course moved back to Bohai mouth. The migration of the people from the exposed shelf toward inland had the main access along Huang He and they followed it to its middle course, reaching Qinghai and Ningxia.

The second area was at current mouth of Yangtze and its lower course, so they followed this river upstream course.

An even smaller Neolithic area (the third one) appeared at the Pearl River mouth, spreading north and south along the coastline.

All three Neolithic areas were connected to the largest rivers of China and with the coastline.

Neolithic people of China were mostly farmers and not hunter-gatherer. This is an indication that they proceeded with a continuation of the Late Paleolithic/Mesolithic agricultural occupation. By contrast the Jomon people of Japan were hunter-gatherers. On the current mainland the rice cultivation came 8,000 years ago, while the first evidence on cultivation is 13,000 old (in Jiangxi Province, on Yangtze southern shore) and it corroborates well with **exposed shelf cultivation at Yangtze paleo-mouth that was 13,900 years ago. This may show that the first cultivation was at the paleo mouth and spread westward 13,000 years ago or upstream of the Yangtze River**. Even then, the cultivation of rice needed 5,000 years to get widespread on current mainland, outside the Yangtze and Huang He basins. The **main farming area, having the greatest population density, was in the Huang He Basin**.

Again I can speculate that this was so because the main rice culture remain traditionally on the shelf, and current coastline was its western limit (that plain on the shelf was 600-800 km wide between Yangtze and Huang He mouths, around 15,000 years ago).

In the end I would say that there is now evidence that **since 15,000 years ago China had the worldwide oldest farmer population**. This implies the earliest self-domestication process, too.

Here we have a sum of events:

-**optimum climate during the LIA was 38,000-36,000 years ago;**

-first moderns from East-China/Yellow-Seas-shelf crossed to Okinawa (Ryukyu) 34,000 years ago, and they live there in the era 34,000-28,000 years ago; they may have perished during **Aira volcano (M 8.4) eruption 28,000 years ago** (the disruption in population covered the entire epoch until 26,700 years ago, when evidence of inhabitation re-appeared); the same eruption may have had catastrophic consequences for some of the East China Sea Shelf population;

- during the LGM the climate was enough warm to support the development of a large coral reef just south of Okinawa; 22,000-24,000 years ago people from Amur Basin migrated to the Sakhalin-Hondo peninsula refuge;

-rice cultivation on the shelf started 15,000 years ago;

These four pieces of information allow one to speculate that the "**moderns**" populated the "shelf" before 34,000 years ago, so more precisely they **were there during the climate optimum of LIA (38,000-36,000 years ago). They gave rise to a regional, but particular type of nucleus, within human social development (Advanced Paleolithic culture of the East China Sea Continental Shelf) that gradually transformed, and around 15,000 years ago, or even before, it became a "farmers' society"**.

Also the arguments about this **rise on the shelf of a local (regional) "culture"** are also the arguments that indicate that **the subject "culture" gradually perished, due to sea flooding** (in the epoch 10,000-6,000 years ago)**, and it could not disperse significantly inland before its demise**. Right now the only eventual evidence can be the 20,000 years old pottery from a cave in Jiangxi Province and the 18,000 years old pottery from Hunan Province, both placed just south of Yangtze Valley, and the 30,000 years old petroglyph with symbols from the middle Huang He Basin (Ningxia Province).

By contrast the Jomon Culture entered Japan sometime 14,500-12,000 years ago from the same continental shelf that included the Korean Peninsula. The Jomon were hunter-gatherers and turned into sedentary population in Japan only 8,000 years ago. Their oldest pottery is 16,000-15,000 years old. Similar pottery was found in Russia Far East (Primorie),

Manchuria, Korea and China. It is assumed that this pottery was used by the local hunter-gatherer people of the Late Paleolithic. The same Paleolithic people used petroglyphs with symbols.

I insisted in this chapter over some aspects, which are not common knowledge, and are mostly ignored by the mainstream science journals and publications. On the other hand **this East China Sea Shelf (ECSS) may contain the earliest stage of human civilization that, opposed to general farming from everyplace else, continued to preserve a nonlinear approach and thinking mentality.**

17. VISUAL THINKING

However "visual thinking" had its own benefits and rapid diminishing of this ability generated a gap between the old and new world. There is proof that many individuals in late Neolithic and during the entire Antiquity retained such atavist skills, and they became the geniuses of those eras. According to some estimates the visual thinkers still were well over a half of total population before 2,500-3,000 years ago.

However a study (1990) on a group of people from the New York City found that people remember 10% of what they hear, 30% of what they read, and 80% of what they see. This indicated that the visual memory still was the most important part of the total memory, while visual thinking was dominant in only 30% of the individuals of this group study.

In mental processes symmetry organizes visual perception because the forms from the visual field are harmonic representations. According to Attneave, the fundamental property of visual perception seem to be the **information compression**. Straight and smooth curving edges in an image can be compressed to their end points, which become sufficient for recognition. Symmetry and periodicity are the candidates for operating the compression process. The visual cortex compresses and decompresses the images. Neurons fire optimally when a visual edge is present in visual field. The image is the veridical facsimile, or internal copy of visual world. **Attneave's harmonic model explains the various alternate percepts, which form some illusory geometric figures like the circle, square, diamond, etc.**

The real purpose of symmetry is to use that regularity to make predictions about the unseen portions of the image. A single geometric signal, representing a particular type of symmetry, can be interpreted in a compound geometrical shape; it creates the percept of the whole when only a part is exposed.

Now the resonance of the brain with its own past states help explain the memories: the past become present through a kind of resonance with the past itself (this is called "morphic resonance"). Similar patterns from the past **give each species a collective memory**. This can explain well the concept expressed by Karl Jung on archetype and collective memory.

The research has found a biological software that converts electromagnetic waves into sound waves that connects visual with hearing, and explains the mental connection made between seen (visual) and heard (sound) patterns. New technique (software) that make possible the "gamma sonification" can explain this mental bridge in the brain, but also it gives us a good idea about the "gamma-ray-language" of the DNA.

Today "visual thinking" rises a lot of scientific interest because people begins to realize that our "old brain" was capable of extraordinary things, which now are no longer possible. We start to understand that "visual thinking" was developed during at least 2 million years before the emergence of primary speaking. It became the thinking of the hunters and gatherers. It was developed within their game of surviving and reached really very high abilities.

For the ability of visual thinking it has been of special importance the role played by the mirror neurons. This is so because the mirror neurons can identify patterns in the environment. A wonder pattern can be inserted into the mind map of the environment, while the environmental elements will converge toward the subject, altering it until it will fit well into the given environmental conditions. This will be the environment answer to human or hominin question about a particular issue of interest. It also be a nonlinear answer.

Early stages of visual perception have involved extraction of local features because the edges are sufficient to communicate visual forms. It has been a convergent thinking system, practicing with the isolation of one single answer. It has been mainly relaying on right hemisphere that typically deals with geometry (as it has been proven, in some case it can be accessed the own brain geometric structure) and deity (allegedly created by the same internal brain geometry).

The reduction of complex, high dimensional data to comprehensible low dimensional view gave rise to **visual language thinking**. Visual language predated the verbal language. Recently the scientists discovered that the human brain processes data as "dots" and the images are stored as mental maps with different configurations and densities of "dots". The **brain processing produces and stores "dots" maps**. Here the visual dots maps are very different of verbal maps, also stored in the brain.

The expression of visual language was analyzed on various animals and it shown how arisen the coordination between visual input and sound output, gradually leading to advances in verbal communication and the construct of language. Here the scientists are divided because the research on monkey (chimps especially) do not show a sign language, leading to visual language.

Not coincidently the cave painting expresses the world view of the artist as a configuration of "dots". Even the first symbols engraved on rocks were represented by configuration of "dots", while they belonged to H. erectus who existed long before any H. sapiens came around.

Another scientific research has found that the hallucinatory images are a direct representation of neuronal architecture from the region V1 of visual cortex. In short some cave painting, like those from Australia, have been drawings made by artists in hallucination mode, and the drawings are exact copies of their own brain structures (region A1 of visual cortex).

Let see some of **advantages and benefits** listed by today **science for "visual thinking"**:

-first of all it is a **nonlinear way of thinking**; hence the "moderns" were not at all any planar thinkers;
-secondly it helps think in 3D, which the word does not do;
-thirdly the visual memory is enormous, it cannot be damaged so easy and it is extremely accurate; if one remembers, in many movies it is displayed the feature of visual memory, where the subject recorded unconsciously the tiniest detail of a past action that was being observed;
-it is intuitive, hence it does not require training or education;
-it deals with light and hence it developed light-based magnetic sensors able to assist in the ambient with magnetic sense and

to help measure distances on small scales; the old brainers were able to travel long continental distances and to guide themselves toward places previously inhabited by ancestors several generations before that time;

-helped with anticipation of movement and dynamics; as an example, a studied hokey player endowed with this sense was capable to anticipate adversary

movement by as much as 3 seconds; some chess players seem to have a similar or even better sense of anticipation; it is definitely associated to visual thinking;

-helps with clarity of thought because it can mentally transform images into working diagrams; it can picture an entire complexity into one single image;

-it makes short-cuts between thinking and doing because promotes fast decision making; this results from a very long experience with hunting;

-promotes spatial strategies;

-allows association of ideas and options;

-help focusing;

-help with self-reflection and self-monitoring;

-reveal relationship and pattern; hence it develops creativity due to pattern recognition;

-enhances transparency; by comparison the "*word*" of verbal thinking can easy hide a meaning, promoting meaning maneuvering; the ambiguity of words incite to violence;

-the world of "visual thinking" was a world without war (war appeared around 4000 BC);

-assists very well in designing things, allowing extreme accuracy;

-and maybe not as the last, it allows cross culture and cross language communication.

"Visual memory" associated with "visual thinking or intelligence" was very extensive and accurate. Some scientists estimate a quantic mechanism at work behind the brain functioning. There is an unproved yet theory that "visual memory" was inheritable and that can explain the transmission of previous knowledge over very long periods of time before any writing/ recording was in use.

Neuroscience research and studies have focused on neural system working and have tried to explain how visual perception and thinking works.

Some of the studies (like Robert Shapley, 2008) explain: *"Analysis of visual system leads to the surprising conclusion that linearity is a rare and prized commodity in neural signal processing. One reason that nonlinearity in neural information processing is the default is that neural communications is mainly through synaptic transmission, and most synapses are very nonlinear. The retina has very special synapses, the ribbon synapses, and these specialized synapses appear to enable the retina to make use of linearity of signal transmission. The visual cortex seems to adopt a different approach. The visual cortex must deal with the nonlinearity imposed by the spiking mechanism of spiking neurons that feed the visual input. The cortex developed an intricate signal-balancing act to reconstitute a linear visual signal in the cortex. Cortical linearity is not simply the default result of convergence of excitatory inputs but rather requires extensive cortical computations. Thus the* **simplicity and elegance of linear systems are created in the visual system, and presumably also in other sensory pathways**. *At the conclusion of this review paper I will offer some ideas about why the visual system works so hard to create, and then to reconstitute, a linearly filtered version of the visual world."* "*This is because linear systems obey the principle of superposition. Superposition means that the response of the cell to the sum of two stimuli must equal the sum of the responses to the individual stimuli.* **Any realizable input can be decomposed into a sum of pulses** *at different times, at different heights. This process of summation is called convolution."*

However the neuroscience research found more interesting results, when the focus was the **brain rhythms influence on visual system**. J. Martinovic and N.A. Busch (2010) explain: *"Cortical oscillatory activity in the gamma-band range (over 30Hz) is a fundamental mechanism of neuronal coding that arises during a range of cognitive processes in both animals and humans. Since the first report (1989) on high frequency oscillatory synchrony between V1 neurons belonging to the same orientation column, the role of such oscillations in visual perception has been extensively researched. An abundance of experimental evidence now links both evoked and induced high frequency* *oscillations to a range of visual stimulus properties." "These strongly suggest that cortical high frequency synchronizations constitute a neural mechanism that subserves processes essential for the organized intake and analysis of visual information,"*

A group of researchers from Netherlands (T.A. Graff, J. Gross, G. Paterson, T. Rusch, A.T. Sack and G. Thut) in 2013 found: *"These oscillations in visual performance cycle at the typical frequencies of brain rhythms, suggesting that **perception may be closely linked to brain oscillations**. … visual stimulation at 10.6 Hz shows: 1) has specific behavioral consequences, 2) leads to alpha-band oscillations in visual performance measures, 3)correlate in precise frequency across individuals with resting alpha-rhythms recorded over parieto-occipital areas."* This shows underlying periodicity in visual performance and may reveal these frequencies functional role.

Another study (E. Spaak, F.P. de Lange, and O. Jensen) in 2014 suggests that *"**oscillatory brain activity in the alpha band provides a causal mechanism for the temporal organization of visual perception**."*

However how our brain processes "nonlinear reality" that surrounds us is still active in the East Asian thinking and culture. It is a proven cognitive differentiation between the members of Western and Eastern cultures, showing that cognitive styles are connected to local culture.

The Easter Asian mind is good at nonlinear, synthetic and holistic thinking. They look for generality, commonality, sensibility, continuity and larger picture. They believe that future is dynamical changing, and pay less attention to current situation. They go long term and value experience.

Four major differences arises between the Eastern and Western cognitive styles. For the Eastern, the name is "holistic" and/or **synthetic**, and it shows:

-orientation to the context or field as a whole (most of attention is paying to the relationship between a focal object and the field);
-preference explaining/predicting events on the basis of such relationship;
-an approach that relies on experience based knowledge (as opposed to abstract logic and dialectical);
-an emphasis on change, recognition of contradiction, and the need for multiple perspectives.

The Western is **analytical** and it shows:
-detachment of the object from the context;
-tendency to focus on attributes of the object;
-preference for using rules in order to explain and predict an object's behavior;
-decontextualization of structure from content, use of formal logic, avoiding contradiction.

The same cognitive differentiation manifests in the way the Easterners and Westerners view images, read a text on a page, write and use their native language, perceive the surrounding reality, think in general, remember the facts from reality, or the way they use the problem-solving abilities.

This cognitive difference has roots in different ways the brain processes the reality in the Western and Eastern cultures. As it can be observed, the **Eastern culture developed the agriculture ahead of Mesopotamia and Western culture, but it preserved the original nonlinear approach in thinking and processing the reality.**

The East preserved a centripetal approach (synthesis of the context's elements), while the West developed the centrifugal approach (logic analysis of a subject removed from the context).

Why occurred this situation, where the Western mind turned toward linearity and suppressing of the synthesis and of the nonlinear? The answer still comes from the discovery of domestication and agriculture, because both these elements introduced a certain linearity of doing. Even when the Eastern mental frame preserved a nonlinear view of the reality, the using of domestication meant inserting a type of control and linearity in the daily life and in the practice of doing.

Not coincidently the Eastern cultures tried to stay isolated from the exterior linear views (European) and their influences (like in the societies from China and Japan).

18. CYMATICS.

As we have seen in the previous chapter the rhythms in the brain play an essential role in its neurological functioning. I will mention two other research, which may bring light on the topic of this chapter that is Cymatics.

Hence, in Dialogues in Clinical Neuroscience, Gyorgy Buzsaki and Brendon O. Watson wrote an article as Brain Rhythms and Neural Syntax: Implications for Efficient Coding of Cognitive Content and Neuropsychiatric Disease. As the title suggest the **brain rhythms help produce a neural syntax**, and **help produce, analyze and store the cognitive content**.

Another study from MIT (September 2011) bring an important contribution, showing that the **brain rhythms have an important role in learning**.

Now, seeing all these frequency-related-activities in human brain, one must imagine how important is the role of frequencies from an exterior sound source, which can interfere the brain activity and resonate these brain rhythms. I have to mention the naturally circulating rhythmic signals, and the phenomena, like the binaural beats.

Wikipedia says: *"Diffraction refers to various phenomena which occur when a wave encounter an obstacle or a slit. In classical physics, the diffraction phenomenon is described as the interference of waves according to Huygens Fresnel principle. These characteristic behaviors are exhibited when a wave encounters an obstacle that is comparable in size to its wavelength.* ***Similar effects occur*** *when a light wave travels through a medium with a varying refractive index, or* ***when a sound wave travels through a medium with varying acoustic impedance. Diffraction occurs with all waves, including sound waves, water waves, and electromagnetic waves*** *such as visible light, X-rays and radio waves"*. So **Cymatics refers to this phenomenon of diffraction of the spherical sound waves. But similar patterns can be obtained when the light is processed through a medium with varying reflective index**, like in the air of the atmosphere, or like inside the fluid of underground magmas.

Everything in nature vibrates and produces sound waves; sound bring information from the environment, not just what can be seen, but from far away and from behind objects. Vision and sound are very intertwined, and some people associate sound with color. On the other hand, rhythm is a language of its own that produces periodic sound vibrations. From rhythm we have the music as an exclusive sound phenomenon. Even our spoken languages produce sound phenomena, and particular rhythms. All these sound waves and the patterns they form, during their interference with a medium (being it sand, air, magma, water, etc.), are the very subject of the Cymatics.

Cymatics is the name given by Hans Jenny (in 1970s), a Swiss doctor, to a new domain of research and the research techniques, which make sound visible. The sound waves, interacting with water (as diffraction), reveal symmetric patterns with 2 D geometry out of the spherical (3D) nature (eventual holographic nature) of the sound. When a surface is under the influence of vibration, it distorts in a non-uniform way, radiating outwards from the center series of patterns, because matter is geometry in motion.

As one of these techniques (the CymaScope) revels, it *"transcribes the periodicities in a given sound to periodic wavelets on the surface and sub-surface of water. The structures created by sound's imprint on water are quasi-3D in nature and may be considered analogs of the vibrational data within the sound. The images presents a slice (2D) through the spherical propagation of that particular sound and approximate the gentle arcs in the dynamics, revealing ratios that can be used to create 3D geometric models"*. *"In CymaGlyphs-the name given to a sound image- you will see a coherent relationship between the sound of a star, a single cell, a human heart beat and the harmonious sounds from a musical instrument. Many contain identical ratios and demonstrate that the laws of Nature are as just as much at work in the heart of a star as within a single cell, a human heart or within heart-felt music"*. (*Clay Taylor: "Holistic analysis of CymaGlyphs reveals the fundamental rules and behaviors involved in giving shape to these portraits of energy"). "Dr. Stuart Hammeroff, physician and professor of consciousness at University of Arizona, U.S., states: Knowing there's this interconnectedness of the universe, that we are all interconnected and that we are connected to the universe at its fundamental level, I think is a good an explanation for spirituality as there is".*

I took the liberty to quote from "cymascope" site on the Internet because these explanations on their technique are a good example for anyone who try to figure out the **direct influence of sound on the rhythms of human brain**.

However in October 2014 it took place in Germany the First World Cymatics Congress, as an integrative approach to science of sound as a creative force.

In this chapter I will try a brief encounter with cymatics; cymatics have a very large presentation on the Internet, so it will be useless to repeat what others already said.

"Visual thinking", combined with sound perception, were at the origin of early discoveries and later experiments in *cymatics*, which have led to early knowledge on geometry, on arithmetic, on symbols and the primitive startup of the incipient technologies of Neolithic and Antiquity.

"Visual thinking" and acoustic abilities also generated the use of symbols because the sound of vowels or some words pronounced inside the caves with special acoustics will leave on the fine sand or any other fine material of the floor some specific geometries. They found out from their speaking sound (of vowels) the patterns (left on sand) with geometric forms, which suggested the way to write down the spoken language.

The path from symbols to writing was not too long; the sacred languages (Hindu, Hebrew, Mesopotamian, Egyptian and Chinese) used the direct copying of the geometry of spoken words as imprinted on the sand floor of the caves to make the pictographic type of writing.

The extreme accuracy of the megaliths can be explained also by employing the "visual thinking and memory". It is said that the Neolithic people discovered by accident how to make artificial stone by turning clay into stone. There is to believe that most of ancient constructions, which seem impossible to construct with very primitive technology and tools, were in fact built in very creative but simple ways.

The same "visual thinking" contributed in part, much earlier in pre-history, to the rock and cave painting. It is assumed that the ingestion of particular plants have produced hallucination, which disclosed or mirrored the design of one's own brain networking structures (like in the caves from Australia).

Even at an earlier time (300,000-250,000 years ago) in Central India appeared the pictographs with dots (the dots-glyphs). Today we know that the dots are recorded by the brain out of perceived images, and **the dot-cluster is the element stored in the memory.**

The frequency domain from the image has connection with corresponding neurons, which are resonated and hence excited. Since the incipient language emerged sometimes 80,000 to 40,000 years ago (or much earlier according to some sources), the addition of verbal frequency has enlarged rapidly the brain (because the sound frequency become converted as image frequency); few thousand years later it produced (15,000-10,000 years ago) drastic diminishing in the brain volume; it is estimated that gradual reduction in "visual thinking and storage of visual information" occurred because the brain mechanism appeared now to be of a much smaller size, and the under construction new mechanism of sound signal neuronal recording overlaps and suppresses parts of the visual system.

There is no doubt that the hearing of hunters and gatherers was in range with that of the animals they hunt. On the other hand, the primitive individuals can sense the frequency of plants and then they can make fast distinction between the subjects. This process of smell differentiation must have its own efficiency in order to assure enough food in a reasonable amount of time. Therefore the smell sense was probably way into ultrasound domain, which is where most plants can be distinguished by their frequency (or smell).

One can ask, how the cymatics were connected to the Paleolithic-Mesolithic nonlinear thinking? They first saw the geometry, resulting from diffraction, and later they associated the geometry with function and information, while later discovering the fundamental ratios employed by nature.

Nonlinearity of thought was an evaluation of patterns. These **patterns display structural plasticity that made them vary according to changes**

in context. Their structure contain information, and information generates function. Because information is complex, but also plastic, or adaptive, the pattern can have multiple functions, or the **pattern (or the content) will adapt to changes in the context**, **meaning that the information changes its meaning due to changes in context**.

The same message of adaptation of meanings we can find in genetics: environmental changes are changes in the context and they imply changes in meaning and function.

So the Paleolithic men learned that **the content was adjusted to context**, and thus the variations in the sound frequency (as context) were responsible for several observable phenomena (in content), like cavitation of air and water, transmutation of elemental substances (metals like nickel into coper) during cavitation, transformation of clay into stone, maneuvering of lodestone (magnetite) effects (producing magnetic effects), and probably some other things that we have not discovered, yet.

Maybe **the most important endowment of nonlinear thinking was the randomly developed capacity to read an unseen function out of context**; this capacity was allegedly present in some human subjects at the end of the Neolithic and Early Antiquity. This explains the knowledge and practice of acupuncture and chakras, astrology, philosophy like that inserted in Vedic Hinduism and Taoism, and some other but not well understood knowledge and exhibits from those eras. It was meditation and martial arts, which both stimulated function during silencing of linear reasoning. They learned that practice stimulate function by the means of adaptation.

The Chinese were able to read the function of what is named today the Primo (or Bonghan) Vascular System, and they named it the six (6) essential energy (Qi) meridians of the Chinese Traditional Medicine. Also they discovered the low resistance points on human skin, which were named the acupuncture points, and which were related with the function of the meridians. Coincidently the Chinese meridians correspond to 6-7 chakras of the Hindu Tradition. These functions were produced by the specific vibrations, or pulses of energy present in these systems.

All this probably allowed an **early maneuvering of function for some elemental substances**, which were associated with the use of the **function**

represented by particular proportions (elemental to nature), and which even today represents an extreme technology and a deep mystery within.

However cymatics was the primordial stage, allowing into visualization of function.

19. "Generative Theory of Shape" of Michael Leyton.

For the scope of this book the theory of Michael Leyton has a very special importance. In my opinion this theory is capable to **explain the works of human brain as visual intelligence, and it can also explain the basics of all natural designs created by natural programs.** Not coincidentally I placed this chapter after the Cymatics because they are deeply related.

My comments in this chapter follow Stephen R. Wassell review of M. Leyton's book from 2001. I have not read the book because it will imply a too high level of mathematics; but the review has had enough information for me to understand the whole idea and to give to it a very special though.

In short *"Generative Theory of Shape"* indicates that *"shapes are viewed originating from primitive shapes (e.g., circles, squares, cubes, cylinders, etc.), primitives themselves are ultimately defined from the fundamental building blocks (point, line, plane, etc.) and shapes are mathematically specified using group theory. …the information necessary to generate a given shape is recoverable from the mathematical definition of the shape. His theory works by capturing the evolution of the shape from the building blocks, allowing for complete mathematical specification of quite complex shapes."*

"The hallmarks of Leyton's theory are transfer and recoverability. The ability to **"transfer actions used in previous situations to new situations"***; the generative operations must be recoverable from the data set, from the end result of the generative process".* **This part of the theory indicates how the design can be stored as information, and how can be transferred from one work to another.** Leyton explains: *"there is an equivalence between the concept of geometry and the concept of memory-storage. …a geometric object is*

defined as a structure from which one can recover the sequence of actions that created its current state".

Even if his book is limited to certain fields of mathematics, he makes considerations on art, music, physics, architecture, biology, etc.

Comparing M. Leyton line of thinking with the general idea of cymatics I found a striking connection that consists into **a natural program capable to generate the shapes of our world by the way of sound and light vibrations** (frequencies). This means different frequencies of sound generate shapes from simple to complex. In the meantime our brain would have the capacity to analyze and store this geometric information by the abilities of "visual thinking". Now such a mental ability will have the capacity to *"recover the sequence of actions that created its current state",* and this can occur from live images or from stored images, **all images being a sum of geometries.**

Now we have the image of, hidden until now, the process that has been the greatest mystery about the H. sapiens vision thinking: **the analysis of the shapes from the environment can manipulate such information content to obtain interpolations and comparisons, extract simple patterns out of complex one; the analysis can extract historical information direct from very shape of the subject.** Complex geometries can be understood by the same generative process.

All these geometric-analysis possibilities depend on proper neural circuitry. As long as our brain can process information explained by the cymatics approach, a similar process of geometry analysis can deal with the rest of visual subjects of the environment.

The same mechanism, being explained by M. Leyton, has been at work in our primitive art and music many thousands of years ago. Mathematics is a language designed to convey information. The same conveying occurs in music and art, which have their own building blocks and patterns; while it can be produced from sound particular harmonies, which are employed to create the geometry of various elements of the environment, the same type of conveying and building of harmonies operates in art. This is why the sound of the music can be associated with colors and shapes, which are specific for art.

Even if the approach is linear in essence, it is able to find the building block of linearity as geometry. Here we have to add the nonlinearity of function.

Everything can be transformed into everything else because this is the rule of nature. In science, there is a proof that every element from Mendeleev Table can be transmuted (transformed) into any other element on the same table. At present the scientists doing experiments with cavitation are able to transmute nickel into copper, oxygen into hydrogen, but also by sonoluminiscence the scientists can turn sound back into light.

During our brain evolution the nature converted the light of vision into the sound of the language. Now it seems that the **nature is in the process of converting back the sound of the language into the light of the vision, and in doing this our type of intelligence will switch, too, because the nature employs a type of generative law that has geometric fundament. This geometry is information and function, becoming by diffraction a geometric expression of energy.**

Our building blocks, the DNAs, convey information into our body and brain, and then back into the environment to reconnect us with the outside world. By doing so, the DNA employs the same **generative rule** for shape, mode or sense that is like a "**universal language" capable to auto-translate everything into everything else.** This is like a primordial language able to understand every other language that ever evolved from it. This is also possible because the **DNA contains the basic generative structure** that permits to every complex structure to evolve out of this basic into whatever shape the interferences will lead to.

We will see this in the next chapter.

20. TALKING TO THE DNA.

Since we started to speak, or even before that time, the issue of DNA communication had been important in human evolution.

In 1990 Jeffrey Delrow discovered 4 letters of the genetic alphabet in the DNA, which form a particular fractal structure. Later the same particular fractal structure was found in the human speech. It was advanced the idea that **human speech might resulted from a DNA pre-programing**.

Numerous experiments were made on the influence of music on plants and animals, and it was discovered that both plants and animals coincidently react favorable to the same type of music that generated maximal absorption into human DNA. The conclusion was that **the DNA, in all living organisms, "speaks" a "universal wave language" that can be decoded or encoded by each species**. This language was resonated by the sound of music.

The science research had proved that human body emits and receive photons. It was discovered the presence of particular wave frequencies, which are used by the DNA to do its work. These frequencies are in the ultraviolet domain of radiation, and for example, a frequency of 380 nanometers have a very specific role, like it allows the carcinogen cells to absorb this frequency, and then to scramble it, modifying it before to re-emit to other cells. Non-carcinogen cells are penetrated by this radiation without interference. This work was pioneered in 1970 by Prof. Dr. Fritz Albert Pop biophysicist at the University of Marburg (Germany). Many other scientists have done extensive work on photon emission and reception by the DNA, including the experiments in 2000 at the University of California and in the previous research done in Japan.

We know today that our DNA processing works in a manner similar to the grammar (syntax and syllables) of any human language. Because of this we can theoretically address our own DNA because the DNA will translate everything into its own language (probably by a microscopic-quantic process). This made the modern scientists curious to make speaking experiments with the DNA.

This knowledge, being inscribed into old myths, seems to exist in human culture since or even before the Neolithic: talking to the self, to animals and plants was considered to be the seers and/or shamans tradition of the old ages.

For our subject that is human brain evolution, we can consider this practice of talking (voluntarily or involuntarily) with DNA as certain evolutionary feature. It has very old roots in the practice of shamans, where the shaman uses the music-rhythm and incantation to produce his magic. Hence this aspect is connected with the advent of music and speaking: the **DNA has selected the specific sounds for composing the music and the most coherent sounds for composing the language**.

Building the contents of the music and/or the language has been again a generative shape processing that has combined the primitive building blocks of sound into innumerable complex variants with the scope to produce and express enough diversity to fit the particularities of the human individual.

A function is a set of rules. In biological systems two different forms of the same protein can have different, even opposing functions, depending on the tissue (that is the context). The rules are the input of information, the function is the output. The way the building blocks of information are combine generate the geometry of the pattern. The plasticity of the pattern implies the breaking and changing of geometry that results from changes in the context.

In this case the music is the chain of harmonics that has a **generative (composing) role** for adaptive structures, like the biological structures. By comparison analysis has an extracting role that dismantle the whole, and ruining the overall meaning that cannot be recomposed in the same settings.

The harmonics are similar with fractal structures. Hence music can be seen as a composition of fractal structures self-similar with the biological structures. These harmonics are present in the DNA functioning frequencies and within all biological rhythms.

Geometry is the workings of the natural programs. Its rules are the symmetry rules. These rules show plasticity, allowing the breaking of one symmetry and the composition of new symmetry. The symmetry depends on the context; changes of context changes the symmetry.

I would say that **speaking itself represented a sound translation of the DNA wave activity**. Maybe sometimes in the future, we will discover the basic connection between UV radiation frequencies and our languages modulated frequencies.

An experiment did by Glen Rein with Quantum Biology Lab in New York in 1998 found that **music can resonate with human DNA** (placed in some tubes). For example the classical music caused 1.1% increase in absorption, rock music 1.8% decrease in absorption, Gregorian and Sanskrit chants cause a 5.0% to 9.1% increases. It resulted that some stress-induced genes might be switched on under sound simulation, while the level of transcription increases. Hence **genes can be switched on and off by sound simulations**.

Richard J. Saykally from the UC Berkeley found that adequately hydrated DNA hold for greater energy potentials than dehydrated strands. He found that slight reduction of energized water, bathing the genetic matrices, causes the DNA to fail energetically.

Lee Lorenzen and Steve Chemiski found that six-sided, crystal-shaped, clustered water molecule that supports the DNA double helix, vibrate at specific resonant frequency, like 528 cycles/second, and support the healthy matrix of DNA. Any alteration of normal hydration can affect every physiological function.

We already know from cymatics that the geometric forms can be created by sound from the simplest to the most complex one, like the complexity of the fractal geometry. Basically the first symbols have the origin in the geometric expression of sound. The geometry and all mathematics have the same common origin in sound frequencies and generative shape.

However, light transforms into sound and vice-versa. So, what is the frequency of the sound as transformed from the UV radiation? The phenomenon of cavitation produces an implosion by sound (wave) energy accumulation into the water molecule, which imploding transforms the sound accumulated energy back into light (sonoluminiscence).

Let's return to our subject of human brain evolution. As it is estimated today, this practice of verbal education appears to be at the very origin of **self-domestication**; it allowed H. sapiens to verbally educate the youngsters, but also to model their behavior. The result was an **accumulation of knowledge that went from one generation to another (that ultimately is an accumulation of intelligence) and a new form of cumulative culture**, which generated a sharp distinction between H. sapiens and all other members of the ecosystem.

As the language rapidly progressed, the accumulation of intelligence progressed in the same rhythm; it was a serial accumulation that was linear because the availability of brain's structures to be dedicated to verbal processing was very limited; it was no place for parallel processing, like in the visual processing system. Previous abilities of visual processing evolved unconstrained, or without competition, and during a very long period of time (maybe 2 million years or so).

Now the new verbal processing ignited from scratch and took over whatever was possible to catch. The **visual had parallel circuits and parallel processing and many other structural advantages (like neuronal networks) developed during its very long evolution**. In my opinion, **verbal ability was forced into linear, or serial processing, because it was constrained by an issue of immediate neural availability and circuitry shortage** inside the brain's space. It also evolved within a relatively short window of time (maybe 30,000-20,000 years).

The 1990 Nobel Laureates in medicine shown that **primary function of the DNA is** not the protein synthesis but the realm of **bioacoustics and bioelectric signals**. All aforementioned experiments of this chapter dealt with the modulations performed by the DNA wave expression.

I would say that the entire civilization we have today originated into our conscious and/or unconscious wave-communication with our own DNA. **The genes are pieces of DNA, and all genetic mutations produced an adaptation of the language spoken by the DNA; new verbal and associated mental abilities of human speech, developed over the course of time, were the copying of changes (mutations and selections) in the DNA language**. This adaptation arises from new set of instructions placed into the gene.

If our languages linking to the DNA maintains, this would show a language evolution under direct influence, or as an interference, of DNA and its self-organization (adaptability) processes, but mostly as a generative processing that assured its large variability and differentiation.

21. The sound of harmony.

Wikipedia gives the following meanings for harmony: the combination of simultaneously sounded musical notes to produce chords and chord progressions, having a pleasing effect; synonyms: agreement, concord, amity, fellowship, cooperation, understanding, consensus, unity, rapport, etc.

In Mesolithic and early Neolithic eras (10,000-6,000 years ago), or even before, people discovered the geometric shapes made by sound and ultrasound.

During the classical antiquity of Greece, the philosophers, who discovered the arithmetic, were thinking that only the prime and whole numbers are important; they played with numerical values assigned to the parts of the geometric figures and discovered special relationship between parts; hence they compared the proportions of the manmade forms to natural forms and discovered that only certain "proportions" can survive in nature; **such particular proportions were named sacred because they show "harmony"**; these special shapes were the spiral, the circle and sphere, the square, and so on.

They have seen in nature the symmetry of shapes that has been translated into the meaning of "harmony". In those early times of our civilization the mainstream thinking becomes increasingly linear, but the combination of visual and verbal thinking have been instrumental in **developing applications for the symmetry seen in nature**. In fact seeing "symmetry" has been the most important part of the **old visual thinking that has worked by recognizing the symmetric patterns within the landscape**. This ability has given the sense of balance and harmony.

Since the advent and progress of spoken languages, human thinking progressively changed, but for the beginning it combined the wisdom of the old visual with new verbal while linear comer. Hence the concept of "harmony" appeared as a parallel to the otherwise linear approach. The **gradual deterioration of this concept of harmony can be quantified by the increasing social inequalities**, which have culminated with violent behaviors, leading to the institutionalization of aggressive conquest and the establishment of the war business (starting 4,000 years ago in Mesopotamia). Even then the **concept of harmony has been maintained for the religious purposes (temples) and for the benefit of the elites (health and medicine)**.

The ancient Greeks considered that the quantity matters only when it has to assure that balance that is harmony; the Greek philosophers were searching for the balance of things and harmonic outcomes. The most important thing in that era was the "proportion" (ratio). The quantity in the composition was not important: the only important thing was the right proportion. In general, in medical use the compounds were small, while they have to preserve the so call "magic" proportion. Even distances were analyzed as proportions between the elements of the environment.

The ancients discovered the harmony of music because they understood the *proportionality present within the rhythm*; practically they discovered the frequency by the means of rhythm; and next they discovered the means of resonance. Harmony was inserted in their architecture, in their temples and into everything else; the **idea that the nature is in permanent harmony that is the harmony of proportions, and it deeply dominated the Greek culture.**

Not only the Greeks but most of ancient civilizations (like Mesopotamians, Persians, Egyptians, Phoenicians, Hindu, Chinese, etc.) drawn and constructed everywhere representations of natural winners of resonance, depicted harmony (flowers, plants, spirals, spheres, etc.), praised the "magic" of proportions, and understood the subject of resonance along its life and death role.

All Chinese Traditional Medicine was and is based on **"magic" proportions**. The ancient practice of acupuncture was based on providing balance to human body by opening clogged channels of energy.

The Chinese and Hindu still believe today in the deep *meaning of harmony and its connection to proportionality*. Harmony was a concept in

5th to 4th millennia BC in the Balkans (Starcevo, Vinca-Cris-Cucuteni-Trypolie cultures), and no war making was in use.

All cultures of Antiquity were dealing with small amounts of "magic" substances and the miracle came from the exact but mysterious proportions. The same principle was applied to the use of magnetism as well as the use of natural radiation, where always the small amounts were the most praised, being the most proficient.

And all these indicate a previously-developed **nonlinear culture**, as it has emerged from "visual thinking". Visual expression is a form to relate to the world, being an outcome of representational process. **Visual expression is a form of action** (artificially simulated words), **and a matrix of realities within realities**.

In Physics coherence resonance is defined as an addition of certain amount of noise in excitable nonlinear systems, which make the system's oscillations to respond in the most coherent manner. In the meantime, the resonances are regarded as excited states of more stable particles. In order to cause resonance, the phase of the sinusoidal wave, after a round trip inside a cavity, has to be equal to the initial phase, so the wave will reinforce, causing wave amplification.

Depending on the situation, the amplification can cause violent swaying motions and catastrophic failure, destroying a particular structure. In natural reality this phenomenon of resonance provides trial for all systems; sometimes it constructs, sometimes it annihilates. The Neolithic people have learnt to understand the cases of natural success and they depicted them in their art, construction and everything else, where they inserted the meaning of balance and harmony.

It has been an early Greek (Hellenic) and Cycladic predilection for harmony of parts and conscious preservation of proportion. In Starcevo and Vinca cultures (7500-5500 BC), **symmetry was used as a fundamental principle**, which places patterns in organized compositions, the harmony of parts in one entity. The expression of the geometrical approach implies that the **Neolithic community developed a strict psychological definition of space and its organization**. This Neolithic case presented a contrasting reality of thinking patterns compared to previous era from Late Paleolithic and Mesolithic (or before 8,000 years ago).

Neolithic communities (since at least 7,500 years ago) developed rational perception of space, and employed its structural manifestation in the production of material culture and the ideas incorporated within (rational thinking). **Analysis of the Balkan pottery art has revealed a linear way of thinking, while it was based on a type of Euclidean geometry (planar geometry). This indicates that at that time the human communities already expressed themselves with a strong linear thinking**.

Since we discovered the complexity and nonlinearity everywhere around us, we begun to understand that harmony means the coherence that defines the complexity. In fact **those magic proportions were creating coherence of the outcome, or we can say the sustainable outcome of the coherence**. Hence magic proportions, balance, harmony, coherence and sustainability, all define one single thing: the microscopic relationship that creates the macroscopic coherence.

However the mainstream social thinking of that Neolithic era was already strongly linear, while the magic of harmony was the domain of few nonlinear thinkers.

In the meantime the practices used to create ancient harmonies help us understand the complexity because the ancients were producing "magic substances", which were coherent complexities. The nonlinear ancients, playing with their "magic", were the architects of complexity.

***

However here comes something that I understood from reading a mathematics paper that was on the discussion site of the E8 group. As appears, the stationary state of elemental particles is the stationary state of quantum mechanics. Now in quantum mechanics the exterior orbits, when in a stationary state, are defined by irrational numbers, like golden ratios (pi and phi). More irrational the winding number, more stable is the orbit. All golden ratios are irrational. By KNM theorem, dynamics with two frequencies in the golden ratio maximally resist perturbation, forming an irrational ratio that exists between order and chaos.

What is the meaning of this?

In my own interpretation, within our linear thinking we find proportions between linear geometry's elements, which produce some irrational numbers, like pi and phi. The row of irrational numbers is infinite. And **some of the irrational numbers (golden ratios) are at the fundament of stable outcomes, which obviously would be the choice selected by nature for all natural products**, which all would enjoy stable orbits for their electrons. The selection begins from the atomic fundament 9stable orbits) and goes all the way into molecular and macromolecular world because at the fundament are the electrons on the orbits, while at the top is the symmetry of the macromolecules.

On the other hand, I see **these irrational numbers as the building blocks of the real world. I think that pi and phi are the real mathematical numbers along an infinity of them, which all belong to the nonlinear mathematics. It should be noted that the same irrational numbers generate the most pleasing symmetries of our macromolecular world**.

As it seems, the people from Antiquity tried to define the meaning of symmetry, and in the process they found the proportions leading to irrational numbers. As I said, these irrational numbers defined the stability at atomic level and represented the true nonlinear fundament of universal mathematics. Our mathematics therefore are valid only and exclusively within the linear layer of reality we created.

Hence the nature, the universe, works with something that we call irrational numbers. But these numbers are universal constants employed in all natural programs. They are present from wave curvature, to standing waves, vorticity, development of plant and biological systems to the shape of galaxies, stellar systems orbits, etc. The golden ratio produces stability of orbits at atomic level that is the stability that generates the symmetry of the macrocosm.

The fractals are self-similar images of the dynamic complex systems, which are infinitely complex. The harmonics show the emerging of each one of them from the phi eigenvector in nested golden sections, like a fractal. The implosion is a Fibonacci spiral converging to phi.

In a paper by Mihai V. Putz (September 2012) it is discussed the Valence Atom with Bohmian Quantum Potential: the Golden Ratio Approach, the author considers that the quantum potential related optimized energy must replace the orbital optimization one. The author has searched for *"quantum valence charge for which the valence energy approaches its optimum value (or the ground state of the atomic chemical-reactivity that is the golden ratio quantification of the valence atomic state)*". He found that *"fractional values in general and those related to the golden ration in particular, may be interpreted as a consistent manifestation of the quantum mechanical (i.e. wave functional) approach of chemical phenomena, here at reactivity level, and may be of considerable utility in refining inorganic chemistry structure-reactivity analysis."* It is here *"a self-release limit that connects with golden ratio by the golden-spiral optimization of bond-order; more subtle, it connects also with 4-pi symmetry of two spherical valence atoms making a chemical bond; such spinning reminds of the graviton symmetry (the highest spherical symmetry in nature)"*.

Avery A. Morton with MIT approximated the positions of the elements in the periodic table with Fibonacci series, and found that the pattern for a Fibonacci series is evident.

These both aforementioned papers shown that the golden ratio has a fundamental importance in generating the macromolecular chemistry of our world, and is the driver of all chemical reactions.Therefore, the golden ratio and Fibonacci series are the generating elements of our molecular world.

It remains to be established the connection between these irrational numbers and natural computation, and also the characteristics of natural nonlinear computation. The result of such advances probably will change our understanding about quantum and macroscopic physics, and our understanding of the universe.

***

The Fibonacci sequence generates the golden ratio, and the golden ration generates fractals, which geometrically are between the dimensions (hence similar with irrational numbers, which are between integers). Again our numbers are not the numbers used by this universe, they are an artificiality.

Now we begin to have an idea about what the nonlinear world of the right brain was dealing with. They saw the first geometric figures on the floor of the caves as a product of sound. This is from where originated the prehistoric symbols and the linear geometry. The symbols transformed into writing and mathematics. But numerical part of the mathematics came from ten fingers, which made the integers. No irrational was in all this. Only the analysis of **symmetry led to irrational numbers because they produced the macromolecular symmetry**. But until today the irrational numbers remained a mystery, like all other nonlinear attributes.

All numbers are bits of information, but the irrational numbers require an infinite information in order to be known. Somehow the irrational numbers are like virtual points, or like limits.

Irrational numbers are the fabric of the universe, while the integers are unnatural but human production. Hence the universe does not contain anything integral (integer) because well demarcated objects do not exist in the universe, and therefore they are exclusive human creation as part of the Ersatz (planetary artificiality created by human mind).

One can fairly guess that here would be an uncountable infinite series of irrational numbers responsible for natural stability and beauty. It is a matter of time until we can translate other atomic and subatomic (quantic) physical characteristics into a comprehensive irrational display. And then we will have a big chunk of universal nonlinear mathematics, and eventually we will be able to do some nonlinear computation.

All numbers, rational or irrational, are ratios between non-integral entities, and as ratios they express a form of harmony.

22. Cave painting and rock art.

One will wonder why we are back to cave and rock art after we reached the point of domestication and the beginning of "civilization"?

The scope of this paper was to find the moment **when pre-historic humans started a direct march toward "civilization"**. Now we have all other information on homo evolution, so we should be able to answer this question.

In many ways this moment is recorded archeologically as the advent of rock art and cave painting.

The evidence shown several stages toward the "art":

-Bhimbeka and Daraki-Chattan Cupulets (India) at 290,000 years ago;

-Venus of Berakhat Ram (Israel) at 230, 000 years ago;

-Venus of Tan-Tan (Morocco) 200,000 years ago.

These three art expressions shown a startup process that failed to progress (evolve) and eventually collapsed. We can speculate that these people were not supported by what was needed to evolve further than drawing some dots, while connecting them into well-expressed shapes failed. It may be the reason that this advent went into a collapse. Now we have the arguments from the Generative Theory of Shape, which show an incomplete development in the brain that is the reason for which the humanoids failed to produce the engraving of shape. This era of the drawing ability failure is recorded as a 200,000-400,000 years old event.

This was a time before the archaic H. sapiens and some other genus homo were the much earlier artists. In the meantime this phenomenon occurred almost simultaneously on two continents: Asia and Africa. In Asia we may have in the Middle East (Israel) some Neanderthals, but thousands of kilometers away in India we can have only some H. erectus or some other local homo species.

The aforementioned engraving from India to Europe, but also in many other places of the world (basically in every old culture), is called "**cupules art" and it appears as the first developed tradition of symbolism**. It has very old antiquity or it is created quite recent; the Stone Age people of Australia, New Guinea and Amazonia still make today the same **engraving of cupules**.

Some scientists see cupules as a symbol but not necessary an art; the **cupules** can have a utilitarian meaning. A very similar case is the **breads and pendants**, which can similarly show a very old antiquity.

Up to this point I can conclude that the abstraction has existed in incipient forms very long time ago, and very much alike it was the result of some very old mutations. Only the mutations occurring in the last 75,000 years have triggered a more complex result that we call today "art", but which is only new mental ability that can connect the cupules to give rise to shapes. This new ability appears in the following sites.

Next are the expressions dated some 150,000 years later in South Africa:

-Blombos Cave engravings (South Africa) 70,000 years ago;
-Diepkloof Eggshell engravings (South Africa) 60,000 years ago.

These two sites definitely belong to H. sapiens.

But very close to that time (60,000-40,000 years ago), we have a Neanderthal site in Europe as La Ferrassie cave petroglyphs. It shows that in the art domain the Neanderthals at that time were on par with H. sapiens from South Africa. It also indicates that both populations were subject to a similar or identical phenomenon that can be guessed as one or several massive cosmogenic influences, which produced significate neuronal mutations and abilities.

However all these manifestations were discontinued, indicating again an eventual but almost simultaneous collapse in Africa and in Europe. All these manifestations shown that the **"visual thinking" was the sole creator of this art**. Again we may speculate that this art manifestation at 70,000-60,000 years ago was not supported by language and/or by music-rhythm, but here many scientists consider the language in existence since H. erectus time. In my opinion, at that time the number of required mutations was still incomplete and unable to support a steady evolutionary process.

Scientists in the business of art consider that **visual arts are a product of processes of engagement and interaction within a changing material environment that is a changing way in which people relate in the world**. Hence the abstract **symbolic process that took place in the mind is the process of engagement and interaction with the environment**, as opposed to generating symbolic representations.

However engraved symbols may correspond with parallel development of language as another system of symbols; the same case of the system of symbols applies to rhythm-music. In the meantime the system of symbols have been a system of communication.

Later we have the first "art" of H. sapiens in Europe manifested as petroglyphs and cave painting. This phenomenon in Europe started 45,000 years ago with portable art and continued with cave painting (42,000-40,000 years ago). The archeologists are not sure about who did this art: the Neanderthals or the Cro-Magnon?

Recently in Sulawesi (Indonesia) were discovered seven painted caves, which were dated 40,000-35,000 years old. This shown that cave art painting was not an exclusive attribute of Europe of that time.

The Neanderthals have flutes (one found in Slovenia and was 45,000 years old) and probably drums; hence some support for a complex art existed. The scientists who studied the Neanderthal skeletons came to the conclusion that the Neanderthal's body did not have a proper mechanism and organs for speaking. Even then the Neanderthals were able to emit strange sounds, which were probably used for communication between themselves. However it does not matter how strange is the language, it still requires neuronal circuits to process such ability.

We did not have any information on homo ability to communicate through a whistled language, or about other possible tool-assisted communication, like whistles, drums, or any other percussion instruments. However the same lack of evidence for **tool-assisted communication** characterizes all Cro-Magnons archeological sites.

The direct verbal communication between Neanderthals and Cro-Magnons would be very difficult but not impossible. Eventually the lack of these full skills on the Neanderthal side would put them in a subdued position versus the Cro-Magnons.

However the "visual thinking" had hard time to reach into abstractions. The first form of abstraction was developed in the form of symbol and the **first symbol was the dot (cupules)** as it is shown in the first art exhibits created longer than 200,000 years ago. The same dots found again their use 70,000-60,000 years ago.

But 42,000-39,000 years ago the rock painting used extensively the dot technique. Immediately later the cave painting shown the evolution of the abstraction mode, and the dots were used now to delineate the contour; drawing the contour from imagination was based on the accuracy of the "visual thinking" and "visual memory". In my opinion the shapes of animals in caves have been the first association with the meaning of "qualia" or "quality", like meat, fur, power, etc.

The same "qualia" has been discovered in geometric forms, which later become associated or translated into symbols: they have been symbols defining a "quality".

In this beginning of the abstraction process, the words and the sound were both connected through imagery. In random opportunities homo found the sound drawing shapes (geometries) on the floor of the cave. Eventually homo begun to associate particular sounds with particular shapes.

Speaking was developed in singing/incantation and some form of praying. The sound effects in the cave have assured the needed magic. Thus some *caves were transformed from simple housing units into first temples of hunters.* Was the incipient spirituality connected to new speaking abilities? Certainly it was. But some archeologists see the *cave painting* as the direct result of interbreeding as it has produced *a mixture of abilities between*

Neanderthals and Cro-Magnons. Was this mixture of abilities similar in amplitude with later mixture of "visual and verbal thinking"?

At this point we can speculate that around 41,000 years ago, coinciding with Laschamp geomagnetic excursion, the homo acquired almost everything he needed to start the march toward "civilization". This march lasted for 30,000 years, until the process of domestication and settlement started (10,000 years ago). Generally it represented the process of self-domestication with all its stages in which **the visual gradually becomes associated with sound, production of speech, of symbols, writing, technology, and mathematics**.

The moment from 40,000 years ago was already immersed into the self-domestication process and most likely it was the first stage of this lengthy process.

The **introduction of words, which certainly was a new full ability at the beginning of the deglaciation (15,000 years ago), changed the whole thinking mechanism, allowing speed, efficiency and abstraction modes. It made the brain smaller, meaning that less energy was needed for the brain functioning**. It also indicates that not many more new neural circuits were created since that time. It allowed the hunters to consume less time with hunting because they needed less food; this allowed them to spend more time inside temporary settlements, and gradually to move to permanent settlements.

The hunters by their new verbal ability become able to address the DNA of animals, which is seen as an incipient process of domestication. As the experiments indicate the plants are also sensible to human communication and to music. Music was used from old ages to attract and/or calm animals.

As the scientists have found, the process of domestication is produced through Neoteny, where the tame animals show juvenility. It is assumed that self-domestication has similarly occurred for humans as it occurred for animals.

It is interesting to see that domesticated subjects show Neoteny and juvenility. In the case of domesticated animals their skull and body diminishes compared with the wild ones.

Homo brain has evolved as an information processing machine. For maybe 2 million years homo brain developed the "visual thinking" that was nonlinear thinking. This type of thinking lacked "structure", causing an incessant search from where to "start", running from side to side and jumping forward. For this reason our nonlinear mind probably found many "starting points" and used them intermittently. This fits better with the idea of "regional development".

Understanding and handling problems did not need computing, it only needed one to zoom out to see the "big picture". **Understanding was more important than computing and this was the main result of nonlinear-visual thinking**. It led to the idea of qualia and found a way to abstract this in the form of symbols. At that time quantity was irrelevant for homo.

Our mind is governed by the nonlinear dynamics of complex systems, we are a complex system and we are nonlinear in essence. **Nonlinear reality is not computable**. One has multiple experiences and learn to compare one experience with another, selecting a better option.

The same process of experiencing new starting points occurred for homo, and this drove him into a very lengthy but primary beginning, with numerous "starting points" followed by abandons, until the accumulation of abilities (resulting from mutations) and experiences led to a better option. The same process eventually led to a variety of distinct homo types. Archaic (Early) H. sapiens was one of them.

The better option choice was the beginning of the march toward civilization, and most likely combined "visual with verbal and with rhythm-music into a complex thinking". This event (about 40,000 years ago) was marked by cave painting; in the same time this was the beginning of self-domestication.

23. MULTIPLE INTELLIGENCES.

In 1983 Howard Gardner had proposed the theory of multiple intelligence that differentiates it into thinking "modalities". Gardner chose 8-9 abilities: *musical-rhythmic, visual-spatial, verbal-linguistic, logical-mathematical* and 4-5 more (but less important for our subject). As we can see visual-spatial was considered the second in importance.

In would say that the meaning of *"visual intelligence"* becomes a complimentary concept covering for *"**spatial intelligence**"*; "spatial intelligence" is connected with *high sensibility for sound perception and interpretation*. Here it has evolutionary appeared the **homo made connection between light and sound phenomena, which both are neuronally treated in a nonlinear way**. Probably a sense of magnetic perception has also been developed in the pineal gland a very long time ago and become incorporated into "spatial intelligence".

The sum total of all these three elements made the "spatial intelligence" as the most powerful neural tool for scanning and interpreting the complex conditions and the mechanism of the environment.

The processing of *sound element most probably assisted the development of the sense of rhythm*, as it was particularly selected out from some frequency domains. We can speculate that hitting one stone with another, for particular types of rocks, the hitting generates a very distinct sound with musical qualities. The same musical quality of sound can be obtain by using wood or special leaves. The drums started with the use of well dried up skins.

As it appeared the prehistoric people interpreted sound phenomena as supernatural occurrences, so the cave painting was made in response to echoes, because they thought that the echoes were made by the spirits inhabiting those caves. Multiple echoes inside a cave or cavernous space could mimic the sound of a herd of animals. The acoustic measurements shown a significant correspondence between the painted cave sites and places with strongest sound reflection.

Similar sound phenomena can be mimicked inside the structure of stone monuments built on open fields. This sound effect can be produced with only two flutes, which seem to cast acoustic shadows. In special stone structures (being placed on the ground or in the underground), low voices produce eerie, reverberating echoes and so on. In other places the sound has been manipulated with architecture to produce desired sensory effects. In most of such sites it has been found that the influence of sound could alter brain functions, producing special states of mind.

To conclude, the **acoustical phenomena have been culturally important for prehistoric and ancient people**. There is interesting to note that similar achievements have been obtained by cultures separated by large geographic distances and by large laps of time. This means the **manipulation of acoustical phenomenon was another human copying from the inner works of the brain**, and during Antiquity it supported a type of primitive stage in the development of some high end technologies, like several applications of cavitation phenomenon. Again this has evolved separately in various cultures and during different time epochs.

From these early sound manipulation, as the sense of abstract art was better configured in the brain, the sound manipulation turned into that esthetic and pleasing form that became "music".

Komninos considers that the intelligence results from the integration of three other types of intelligence: *inventiveness*, *creativity* and intellectual capital of a community (let's say *local culture*).

Bethune considers that it is about the ability to gasp a changing whole and anticipate the next stage; or to understand the relationship developed inside a fast-changing-array.

Today analysts of IQ consider that verbal skills help produce the highest IQ score. The IQ high score is seen as a problem solving skill that ignores nonlinearity. Hence the **developing of verbal skills represented the rapid change toward producing a civilization**. However I consider that ***the mixing of musical, visual and verbal skills has produced the highest complexity in the brain that has been the breakthrough toward civilization.***

As it appears, *the diminishing of one of these abilities is a negative factor for the whole brain functioning. It may bring to us new type of primitivism.*

What can be the possible case of this cranial diminishing? Here I have to mention the role of diet. A smaller brain can accommodate lower energy diet, and this is exactly what the domestication of plants and animals have generated. The wild plants and animals were giving much higher level of energy than the domesticated ones. **Hence it has been a natural connection between smaller brains, verbal skills and domestication**.

Our agriculture gradually provide diminishing energy for human diet justified by the need to assure the surviving of larger number of individuals. We eat large quantities of food, but we get lesser energy from it. This gradually diminishes the brain working power and our very own intelligence. The visual intelligence is rapidly diminished, while the complexities generated by these new conditions are less capable to solve the problems of the ambient.

Gradually we have lost our membership to the ecosystem: we are not able to understand the complexity of the ecosystem, we cannot recognize its patterns and the relationship occurring with the ecosystem. We can no longer communicate with the ecosystem, while we can no longer communicate with our inner building blocks, like the DNA. Hence we begin to create a **"linear cultural evolution", where we change the environment to adapt to our genetic needs**, **while replacing the benefits of natural selection with human selection**.

In terms used by the science of complexity we actually generate a new type of clustering around and toward the linearity of verbal skills. We have survived the ambient for many thousands of years based on "visual skills". Now we are losing rapidly this ability to understand the works within the environment; the environment of ecosystems is severely altered; we created new but artificial (eco?) system that does not follow natural rules; this is a surrogate not an ecosystem.

I would say, our current folding with verbal thinking blind our "visual intelligence" and obstruct us from seeing the ambient in its natural way. This obstruction manifests in all science we do.

Verbal skills were an advantage 15,000-8,000 years ago, which increased our human complexity as new addition. But in the course of time, this exaggeration of one-way-only development, broke the natural balance and it turned us blindfold inside the ambient because of losing of the "understanding". We were stripped out of our "visual intelligence", which fact removed us from non-linear realm.

We have created the world of words, where very few individuals can achieve the highest power by maneuvering words. The money result from this verbal ability and is just the imprint of words; money bring to the skillful linear individual the benefit of amassing, and in very large quantities, money that equals (verbal) power. We enforce (verbal) power by (verbal) law and all other possible means. We compete instead to collaborate; we destroy the opposites in this process of competing.

All nonlinear technologies, being on the brink of discovery, are disruptive to current social linear setting; therefore the mainstream society has to ignored or even prevent them from being researched and/or constructed.

In all these we turn against ourselves, endangering our very own existence. There is new Dark Age at the horizon that is the primitive age of extreme verbal abilities.

24. Two brains.

A recent book *The Brain of the Denisovan Girl* by John Harris has a very interesting approach to the transformation of "visual thinking" into "verbal thinking".

He states, "*Imagine two distinct brains, one archaic and one modern. The two brains developed in two different evolutionary settings. The archaic human brain is pre-verbal and it was perfect before language dominated thought-for thinking in pictures. The "verbal" human brain is modern, perfected by and in parallel with the development of human language. This seem to have required the **suppression of ancestral visual thinking**. Instances of modern humans who exhibit archaic brain skills are, at least in part, atavisms. So we are walking around with both brains encoded in our chromosomes, archaic and modern, along with two distinct brain building programs. Occasionally, however, **both brain building programs are triggered, and they fight for control**. The outcomes of the conflict, in this hypothesis, could include autism, genius, photographic memory and synesthesia. These two brains, one verbal and one visual, occupy the same space inside the skull. Perhaps they may compete for this space, perhaps they are amalgamated within it.*"

He explains that the lens of the eye produces at its back focal plane a Fourier pattern. The brain itself is a Fourier processor. He said, "*...there was undoubtedly a parallel or anticipatory change in the brain. ...this **required the further modification of visual memory machinery in order to create, in effect, a word processor**. Perhaps the **visual memory for objects was modified to create a memory for sound and then words**. It might be that edge detection in the eye was a technical prototype for the detection of the edges between their (human) words. There is no great distance between the memory process and the thought process. Thinking in words is probably a more compact process than thinking in pictures. New verbal reasoning* *may somehow step on or strongly inhibit the much older and probably more highly evolved ability to think in pictures.* **Verbal memory and reasoning machine was improvised in humans, maybe 50,000 years ago or less, by simply taking over for word processing some core components of the visual memory machine. Brain's visual memory (analog) computer was modified by evolution to recognize and reason in words.** *Incoming sound must be depicted as a Fourier pattern in order to gain access to a neural memory machine originally evolved for thinking in pictures. Sound become images in the frequency domain that is Fourier patterns.* **A machine that was once finely tuned to process pictorial information surrendered a chunk of its picture handling capacity to the processing of words.** *The story calls for a pre-historic brain with the gates left open, constantly receptive to Fourier patterns received from two different sources, the eye and the ear."*

I took the liberty to put together paragraphs from different pages of this book to make the story concise enough to fit this chapter. Other parts of the book are focused on other issues, like the connection between the eye and hearing apparatuses and their historic evolution in primates toward the advent of verbal expression in human language. Other parts are connected to the story of the Denisovan girl and her pre-historic brain.

I have skipped all these parts because in this chapter I would like to enhance something else that is the case of **verbal mechanism suppression over visual mechanism**. In the meantime it is clear how **an analog brain can process by parallel systems the nonlinear information**. The result of this processing is "understanding" that prevailed during the peak of visual thinking abilities, and it still was in working balance with verbal thinking until the suppression process became a dominant, and verbal thinking took over.

From over one page of quotations the conclusion is that today we have left few shards of the old vision thinking, mostly manifested in illnesses and in geniuses. They are the avatars from the old brain.

But let's see how really the brain works, employing now **the perspective of the brain rhythms**. In the case of visual thinking it occurs alpha rhythms desynchronization over the occipital region, while simultaneously it occurs synchronization over the motor cortex. In the case of the motor response to visual stimuli, it occurs the reverse process. In the case of attention, an alpha slow rhythm inhibits the processing of information, and the activity in other regions of the brain becomes more focused. Desynchronization means suppression, and in our cases this **suppression blocks the information processing, and favor a local network of neurons to work on a local band of frequencies**. This process produces a selection.

Now during intense thinking alpha desynchronizes, while the gamma band turns to synchronize. **Individual alpha frequency correlates with left occipital cortex,** but the rhythm itself originates in the more anterior regions, where initially the word recognition first occurred. Verbal stimuli is accompanied by widespread increases in alpha-1 coherence. By contrast for visual words, there is much less influence upon coherence and widespread decreasing in amplitude. Decreases in amplitude are typical for visual simulation. Oscillations of large populations of neurons of the cortical regions are responses to visual target events. Oscillations in the beta (10-45Hz) frequency are common for the motor system. It was found that the neocortex simultaneously generates different rhythms, and also these rhythms can go into transition to other frequencies. For example gamma-40Hz and beta 2-25Hz can combine to generate a third frequency, like beta1-15Hz.

We can say that global synchronous rhythmic activity of large neural assemblies destroys a selective pattern of local oscillation. In order to activate a local oscillation, it is needed the desynchronization of the large rhythm. Hence, in this case global suppresses local selectivity, but when global stops, the local flourishes.

This shows that **the rhythms produce a complex work inside our brain with processes of synchronization and desynchronization, which alternate between different regions and between the hemispheres**.

Latest studies on the origin and evolution of human speech and language indicate that the presence of speech occurred 80,000 years ago before the out of Africa event. However the language (that implies much more sophistication) advent is dated some 50,000 years ago. Hence **significant changes in the functions of the human brain must be considered for the last 50,000 years; it may be the case of changes in the brain rhythms, and in synchronization/desynchronization processes because we definitely had incoming suppressions during this past time interval** (50,000 years).

Now I like to quote another important paper for this subject: the books of Temple Grandin, PhD; this author had a mild form of autism, and she wrote the book: "*Thinking the Way Animals Do: Unique insights from a person with a singular understanding*".

She wrote in her book: "*I have no language-based thoughts at all. My thoughts are in pictures, like videotapes in my mind. When I recall something from my memory, I see only pictures. I learned that there is a whole continuum of thinking styles, from total visual thinkers like me, to totally verbal thinkers. Artists, engineers and good animal trainers are often highly visual thinkers, and accountants, bankers, and people who trade in the futures market tend to be highly verbal thinkers with few pictures in their minds. Most people use a combination of both verbal and visual skills.*" Another of her research shows that "*humans need to interpret their environment, getting rid of the parts that don't fit their narrative explanation and sometimes combining unrelated bits of data. Normal people have an interpreter in their left brain that take all the random, contradictory details of whatever they're doing or remembering at the moment, and smoothes everything out into one coherent story. If there are details that don't fit, a lot of times they get edited out or revised. The right brain picks outlines and ignores details within.*"

Here is the issue of **causal illusion** because causality is only an imagination opposed to the complexity of reality. **Causal illusion help maintain erroneous ideas in the mind, but also it prevents new information from correcting them.** This is so because seeing a large number of people achieve the desired outcome after doing something ineffective this makes the observer to correlate these two things, and this blocks the acquisition of new but conflicting information. Simply exposing people to new information does not help. **Causal illusion is a linear cognitive trap and people are mentally encouraged to build a causal narrative for all erroneous beliefs, which all only suppress any other valuable and correct information.**

In both papers mentioned and quoted here there is nothing about linearity and nonlinearity of those described as verbal or visual thinkers. But this aspect is well established in science, as it was mentioned in the previous chapters. Also research shows that **language suppresses visual memory that is called "overshadowing".** This means **linearity suppresses the nonlinearity.**

Here the genetic role is of paramount importance. It was found **that between old ape and modern humans 23 genes have been modified: 8 genes in the brain and nervous system, from which 4 are for axonal dendritic growth and synaptic transmission; 2 of them are implicated in language disorder. So the synaptic transmission changed in humans.**

Let's see another important thing, like the **differences between a computer and human brain processing.** According to *Chris Chatham* (Science Blogs, March 27, 2007), we have the following 11 differences: "*brains are analogue and computers are digital; the brain uses content-addressable memory vs addressable memory for computers; the **brain is a massively parallel machine** vs modular and serial for computers; processing speed is not fixed in the brain; short-term memory is not like RAM; **no hardware/software distinction can be made with respect to the brain**; synapses are far more complex than electrical logic gates; **in the brain, processing and memory are performed by the same components**; the brain is self-organizing system; in brains the information is embodied in our body; the brain is much, much bigger than any computer.*"

According to Wikipedia: "*An analog computer is a form of computer that uses the continuously changeable aspects of physical phenomena to model the problem being solved. In contrast, digital computers represent varying quantities symbolically, as their numerical value change. So the brain uses continuous values, which cannot be reliable repeated with exact equivalence.*"

Stanford Encyclopedia of Philosophy is saying: "*Dual Coding Theory of Paivio proposes that the human mind operates with two distinct classes of mental representation (or codes), verbal representations and mental images, and that **human memory thus comprises two functionally independent** (although interacting) **systems or stores, verbal memory and image memory**. Imagery potentiates recall of verbal material because when a word evokes an associated image, two separate but linked memory traces are laid down, one in each of the memory stores. Obviously the chances that a memory will be retained and retrieved are greater if it is stored in two distinct functional locations rather than in just one.*"

This theory has resisted to multiple criticism and experiments (since 1971) and it still holds today. Recent research have shown that information comes to human brain in the form of dots and is stored in the same manner.

Not much research has been done on the issue of verbal suppressing visual, but we have the realities of human civilization, which indicate that, in the last 8,000 years or so, everything we make or build has been linear in essence.

A recent research at University of Rochester (2013) has discovered that IQ score is 70% higher in individuals with a brain that suppresses background movement or motion. The higher the IQ score, the lower were the subjects to detect the background motion (they detected fast only the foreground motion). This research comes as another argument about **our modern thinking system that produces linear intelligence by suppressing a large part of visual intelligence**. The research have not involved any verbal abilities; it has simply tested the visual abilities only, and in relationship to the IQ scores.

Another paper "Conservative Left Brain, Liberal Right Brain" by Charles Block is very interesting in describing the functions of right and left brains, as they appear in scientific research and experiments. I will quote again the most important ideas.

"This illustrates how differently the left and right brains behave when they are not influenced by each other, because the left brain maintains an uneasy alliance with right brain. The left brain is a dopamine oriented world unto itself. The number of dopamine receptors in the left brain is higher than in right brain. Dopamine is a key neurotransmitter, and facilitates the functionality of fine motor skills, language, arithmetic, and monosemantic processing. By monosemantic we mean the single-mindedness, or single meaning of a particular memory or stimulus. The polysemantic thinking of the right brain is not very good at producing language and arithmetic."

"The right brain is norepinephrine oriented world. Norepinephrine is another neurotransmitter, not too unlike dopamine. Both act as stimulants to the nervous system. Norepinephrine receptors are very responsive to the external environment. Here the left brain seeks to avoid internal contradictions in its analysis of the environment. It tends to limit the analytical result to a set of predetermined outcomes. The left brain tend to bend the reality to fit the way it wants to process it. There is a certain…pre-determinism. Time-orientation is a natural outcome of the left brain's propensity for pre-determinism."

"By contrast the right brain has a lower propensity for pre-determinism, and is not as inclined to drive the information analysis to a particular outcome.

It exhibits less linear information retrieval processing than the left brain. Right brain can lead to more ambiguous and complex outcomes."

"The difference between the brains are sometimes subtle, both anatomically and functionally."

"The right brain does very well with new and unusual stimuli. The right brain has positioned the left brain to adapt this information more effectively into its structured analytic processes, and even makes changes to the left brain's monosemantics."

"The left brain is more electrically active than the right brain. It draws more current from the ARAS (ascending reticular arousal system). Conversely, the right brain seems to run on its own, passing electrical current all around, not varying much even when in focused analysis."

"The left brain structure can lead to extreme disconnection from the real world. If the left hemisphere becomes hyperactive, it can construct delusions, paranoia, hallucinations, and a closed world into itself."

Left brain is partially deaf to music, while the right brain is partially deaf to words and sentences. Because of this, the left brain is tricked by words and promises, being easy tricked by politicians and verbal concepts.

Finally on this paper of Charles Brack it is explained how the **conservative people have enhanced the left brain, while the liberals have enhanced the right brain. I see this as the perfect connection between linearity of the left brain and the nonlinearity of the right brain.**

A study by Dr. Linda Kreger Silverman indicates that 30% of people uses the visual thinking (right hemisphere), 25% exclusively think in words (left hemisphere), and 45% use both hemispheres. However, it indicates that these abilities can be switched to match the needed task to be performed.

When we solve a nonlinear equation assisted by computer the result is not a formula but a visual shape, a pattern traced by the computer. This new mathematics is the mathematics of patterns, of relationship. Strange attractors and fractals are such patterns, as a visual description of the system's complex dynamics.

Another issue is "speed". The right brain in some people can visit 32 nodes (concepts, symbols) per second, while the left brain can do on average 6-7 words per second as experienced by typical verbal-sequential thinkers.

Beyond speed the **right brain prove to be paramount because it deals with light, sound, rhythm and harmony**, which are all parts of the wave phenomena.

The hemispheres are controlled by the frontal lobes, which are divided also in right and left. However the lobes control by suppression, inhibition and censorship what one hemisphere wants to transmit to the other one. In the meantime these two hemisphere have an alternating activity that is called the ultradian rhythm and has a period of 90 minutes.

In this chapter I have presented a collection of scientific studies which can assist the reader in understanding how our current society is driven by the two brains we have at once in our skulls. In summary, it seems that our society is somehow equally divided into right and left brain thinkers, while the majority of its members use both brains at once.

Since the advent of full extent of language and the development of agriculture in the last 10,000-8,000 years, the design of our civilization has evolved toward linear constructs and the dominance of verbal thinkers and their linear behavior.

Thinking means linking ideas, and **this linking is called reasoning**. Thinking means to choose a "subject", to remove the chosen subject from its natural context (where it freely floats within natural dynamics), to add to it a predicate (giving it a quality, inserting it in a certain flow, while giving a sense to this flow) that creates a "content", like in a text or sentence. This entire operation takes place in the left brain, and it produces "linearity", or a type of sequential organization, where things are organized by particular logics. The same linear process was applied in domestication and agriculture, and since then, in everything else.

In martial arts the practitioner is asked to get a rid of every thought, or to quit thinking. Thus the practitioner fights the best without thinking (linking) and logics. This shows that the logics are an artificial construct, and are opposed to natural laws of complexity and their obvious rules. It also shows that the logics (linking) of the left brain are distant from natural coherence produced by the whole brain working together.

At the origin both **speaking and music were connected by the rhythm** that is a periodic variation, and it characterizes every natural phenomenon. Our right brain perceives the rhythm, but as the capacity for musical **analysis (linking)** increases, the left brain becomes dominant. Non-musicians perceive music with the right brain.

Music and speech are analogues because of the presence of the rhythm. So the rhythm of speech was sensed at the origin by the right brain. Only the later development of **linking** processes moved it into the left brain. The same meaning applies to **"art", where the same rhythm of periodic patterns** was perceived by the right brain.

Every periodic phenomenon from our own body (like breathing, hart rhythm, and so on) is connected to the right brain. Our health ultimately depends on good standing of such body rhythms. Not coincidently the rhythm of music, or a particular reciting can interfere with our body rhythms and can heal.

I have to say that after the creation of music and language it existed an era, when somehow frequently it occurred the synchronization of both brains (hemispheres) that produced the highest coherence and the best intelligence. Maybe it was only a phenomenon that appeared in singular individuals and not in mass. Regarding the aforementioned era there are a plenty of archeological arguments.

The linear thinkers (25%) have come to dominate the middle section (those 45% with both-brains combined ability) and to suppress, where they can, the remaining minority of visual thinkers' (30%). We are almost prisoners of linear thinking, since we cannot escape this mental framework.

Because of this linearity we have globally developed a large social inequality, where every local social structure basically follows some particular brain abilities (being them linear or nonlinear). As a consequence, we have a market oriented society that is linear and in which linear thinkers are the leaders and the main beneficiary.

Let's compare the job orientation given by these two brains: linear thinkers send in the Marines, the nonlinear send in the diplomats; linear thinkers build prisons and increase police forces, the nonlinear set up educational and diversion programs; the linear thinkers want one nation (theirs) to be supreme, the nonlinear values the community of nations working together to solve the world problems; the linear thinkers like clearly-defined-roles, while the nonlinear like extreme flexibility in their roles.

Hence the nonlinear thinkers, as a minority, must adapt to survive into a world where they are less welcomed because they have the potential for troubling the majority with unpopular solutions and ideas.

***

The brain is cross-wired, with the left hemisphere controlling movement on the right side of the body, and the right hemisphere controlling movement on left side of the body. Communication between hemispheres is achieved by a thick bundle of nerve tissue, named the corpus callosum.

Before 1996 were made various studies to determine the condition of asymmetry between the two hemispheres of the brain. In male they found right side larger, while in female they were more like equal. It was found a significant fraction of gray matter versus white matter in the left hemisphere. The studies concluded that the difference is attributed to gray and white ratio not to the density.

However other differences found have been:

-thicker cortical gray matter in left hemisphere;
-left frontal lobe has higher ratio of gray and white matter;
-right frontal lobe is larger (by 13%);
-right hemisphere seems to have more gray and more white matter overall;
-left parietal lobe larger than the right one (even larger in female);
-density of the matter was higher (gray and white) in the left parietal lobe, also showing a deeper folding on the left side;
-temporal lobe (responsible for speech) larger on the right side;
-lesser white matter in the left temporal lobe.

In general these studies found only few differences between the hemispheres, but shown a larger right hemisphere in most fields, with a left hemisphere larger in less fields.

The lopsidedness, arising during evolution, is attributed to adaptability that produced a plasticity of the brain.

A new study published on April 13, 2013 by the Proceedings of Royal Society B, found that are some areas that are bigger in the left hemisphere than in the right hemisphere (author Aida Gomez-Robles). The study found that human brain shows variations among the population, while humans have larger frontal and parietal lobes than chimps.

A theory of asymmetrization of organisms establishes the commending side among twin parts; it defines the left hemisphere as the operative part that is younger. It also defines that the centers for administration of new function first appear in left hemisphere. The same phenomenon occurs for new genes, which may stay for generation in the left hemisphere, where after verification are transmitted (transferred) to the right hemisphere. A similar process occurs between male genome from where after verification the genes are transferred to females.

The study say that left hemisphere will dominate human development during the phases of evolution and stable evolution, when the function is absent.

***

Now we have the aspect of handedness, where, for example, the right hand (used by 90% of human population) is coordinated by the left brain. This left hemisphere is physically larger and more developed (especially Broca's and Wernicke's zones of this hemisphere are up to three times larger than in the right hemisphere). Also the primary motor cortex of this hemisphere is larger and denser, with neural links more tightly connected than in the right hemisphere.

As we have discussed at length, the visual thinking of the right hemisphere was perfected and considerably developed in the course of several million years. Then, why the left hemisphere is more developed because its ability for language appeared quite recently? There is no science answer to this question, yet.

In opposition to the aforementioned aspect of the primordially of the right hemisphere, 90% of people use the right hand that is controlled by the left brain. Then we can assume the right hand use is a time immemorial adaptation for unknown reasons. In support to this case of adaptation come some scientific findings about the left-handed people: they suffer a disproportionate number of migraines, significantly higher number of problems with their thyroid, the same with allergies and stuttering. They have twice as often difficulties with their immune system, and they show 10 times higher disabilities in learning. Thus we can speculate the use of the left hand, coordinated by the right hemisphere, is a special problem of an awkward minority that also have many other biological, or medical issues. Of course their reveled geniality enter in the same domain of the human split evolution (that 10%) in a less travelled direction.

The research also discloses a sex gender differentiation generated by the same two hemispheres, with boys more readily to visual cues than girls, and with girls considerably higher verbal skills.

As it appears, the issue of migraines generated some 10,000 years ago the occurrence of trepanation, which technique was in use until very recently. The skeletons from Neolithic found in various parts of the world have shown a 5-10% occurrence of trepanation. Studies found that increases in the mounting pressure of the cerebrospinal fluid in the brain increases with age (beginning with 30-40 years of age and accentuating at 40-50 years of age) because of the reduction (with age) in the flexibility of the skull to expand to accommodate the blood pressure pulses. These blood pulses show a variation in volume of 15-20% that is the volume variation required for the skull (or an increase in skull volume of 3-6 ml). As a consequence some neurological symptoms may appear. The old writing mentioned trepanation like treatment of foolishness (incl. dementia and Alzheimer). Also this restriction in the expansion of the skull cavitation represents an adaptation limitation to different living situations, which diminish the living capabilities. It is a scientific proof that variation in blood pressure affect the internal pressure of the cerebrospinal fluid.

For example the research in migraine occurrence found that sudden changes in barometric pressure can trigger migraines. Also changes in food diet, sleep pattern, sensory stimuli (like variation in light intensity, cosmogenic and background radiation, and ultrasound exposure), and changes in stress and medication can trigger migraines, while the most migraines seem connected to variation in cerebrospinal fluid pressure. The use of left hemisphere as verbal thinking is significantly higher in women, while the women suffer the most from migraines.

***

The technique of trepanation first appeared around 8,000 years ago on the Mesolithic/Neolithic border, and used stone blades; most of the subjects survived the surgery and healed. This era marked the beginning of the transition toward linearity of thinking.

A study on trepanations in Scandinavia found 47 skeletons with occurred trepanation; these skeletons belonged to various time periods from Neolithic to Medieval era. Those 7 cases from Neolithic referred to subjects found 40-50 years of age and older; the same pattern of age maintains during the Iron Age for another 12 cases; going to medieval times the skeleton subjects turn to belong to individuals 30-40 years of age. This information allows one to observe that apparently the life duration decreased with 10 years during the medieval era compared to the Neolithic and the Iron Age (that corresponds to classical Antiquity).

Greek and Roman written records confirmed that people during classical Antiquity have lived to 60-80 years of age, receiving a monetary award, as pension, for having a certain age. Hence current theories about the life expectancy from Neolithic to classical Antiquity are incorrect and misleading.

However the need for trepanation (also becoming more like a cultural aspect in various regions of the planet)) expanded all the way to nowadays, indicating a continuation of the host of brain diseases produced by the suppressions responsible for linearity.

***

How different were the brain rhythms during Paleolithic? Most of the neuroscientists will say the brain rhythms are quite constant over the entire mammalian range. This is true, but one differentiation still exists in connection with the sleep period.

Many studies indicated that the hunter-gatherers were sleeping a period of 12 hours/day (in 24 hours), and this sleep was divided into two sections separated by a wake period of 1-3 hours. Until the introduction of electric illumination most people around the world followed the same sleep pattern. Modern pattern of sleep is a continuous 8 hours.

Deep sleep is associated with five phases: 1,2,3,4 and REM. These stages progress in a cycle from stage 1 to REM, while the cycle repeat itself several times. 50% of sleep occurs during phase 2, 20% in REM, and 30% in other stages. A complete sleep cycle takes about 70-90 minutes.

During sleep, the direction of interhemispheric flow of information for delta (1-4 Hz) is from right to left, and for beta (16-30Hz) from left to right. During the state of vigilance the direction of flow changes from left to right.

However the research shown that delta waves, specific for deep sleep, when they show right hemisphere dominance, have more activity in women, while these waves intensity decreases during the lifespan (especially in men), with their disappearance after the age of 75.

As we know the brain rhythms are different during sleep versus the wake time. Therefore, because the pattern of sleep was different until very recently, during a 24 hours span the repartition of brain rhythms was different, too. This eventually produced a distinct functioning of human brain before modern times.

***

Hence **human adaptation to changing conditions in the environment participated in specializing the hemispheres, and eventually was the main cause for various suppressions in the brain, driving to that adaptation that produced the outbreak of linearity as the occurring prevalence of the left hemisphere**.

Our linearity in thinking was in debt to the left brain that was the part of the brain that sharply became bigger and more complex not long time ago.

In the end of this chapter I like to suggest that **the left hemisphere probably was working in compliment to the right hemisphere for very long time**. Because of its complimentary role the **bulk of genetic changes** induced by the cosmogenic radiation in the era 80,000-12,000 years ago (due to a large host of geomagnetic excursions and other astrophysical events) **have been directed to the left hemisphere (where they form a pool, waiting verification for admission into the right hemisphere), where they found more available storage space and circuitry than in the working hard right hemisphere** (at that particular time). Thus this conjecture helped the left hemisphere to develop very fast during this short interval of time (70,000 years).

However **the left hemisphere had a natural built in complimentary role that was naturally attuned to linear characteristics (because it was**

in charge with linear functions). Its linear processing had the scope to supplement with the day-to-day linear abilities (functions) the nonlinear expression of the right hemisphere that was the main one in use for visual thinking processing and life-long strategies. The left hemisphere had produced the tools necessary for simple and normal needs, while the tools could be used by the nonlinear strategy of the right brain.

The development of the left hemisphere was stimulated by the storing of such **numerous new mutations, which all only amplified the built-in but linear processing**. Hence **a side show was on the way to turn into the main show. The final triggering factor for linear strategy activation was the Solovki excursion (7,000-4,500 BC).**

This suggestion might explain how the linear turned prevalent in human thinking.

***

To support this opinion I will bring some anthropological arguments.

In every quarter of the world it is the right hand, and not the left, which predominates; actually more than 90% of people today use the right hand as the main hand. During Classical Antiquity in Europe, North Africa, India, or China the situation was similar with today case of right hand domination.

But **it has not have always been the right hand dominance in our prehistory**. Stone Age tools, cave painting and rock art bring evidence of a more even distribution for the use of the hands. Even some studies shown that it was a higher proportion of left hand users during Paleolithic, and Early Neolithic, like it persisted all along of history among tribes in isolation or within the contemporary primitive cultures.

The information from archeology and anthropology indicated that a massive switch over took place during the Bronze Age (3000 BC or 5,000 years ago). During this epoch the right hand have gotten the upper hand.

Undoubtedly the switching over was not an adaptation or not a cultural aspect being spread all over the planet. It must be considered an external mutagenic factor, like the cosmogenic factor, that was active all over the planet, and led to mutations responsible for the switching over. In fact we have the development of the left hemisphere of the brain that also favored the development of language. Further development of the language became the stimulus for further development of the left hemisphere.

When we collaborate the information about the development of the verbal thinking in various domains, **all data overlap for the epoch 7,000-5,000 years ago**. And this is not coincidence.

As it appears, the development of the left hemisphere was gradual, and manifested in sequences: the creation of large settlements, or cities, the startup of agriculture, creation of symbols and writing, or mathematics. Somehow among the last linear developments, it occurred the switching over from the left hand to the right hand.

In my opinion the main scope of the nonlinear strategy was to find the fundamental proportions (ratios) existing within the nature. To be able to maneuver these proportionalities the humans needed a minimal linearity in thinking capable to develop the needed tools (that is the function and this belongs to the left brain).

But the nature took advantage of the human brain structure existing during the Paleolithic, storing all new potential mutations in the space available in the left hemisphere, contributing to a massive development of this hemisphere and enhanced by neuroplasticity; this hemisphere was primarily responsible for linear strategies of thinking, which were still placed in the right balance with the main nonlinear activity and characteristics of the right hemisphere. New linear skills affected further the neuroplasticity by a gradual skills multiplication and complexification. The way we think can regrow brain cells and change the structure and function of the brain; for this today is a solid scientific proof.

The aforementioned situation broke the balance between the hemispheres, bringing up the linearity as the main strategy. **This switching over had generate the entire civilization we have that contrasts the surrounding nonlinear natural reality. Because of this, the linearity is an artificiality with space and time limitation**. It is totally unsustainable and this limits its development in time. It requires an enormous space that is not available on this planet. It disastrously harms the environments of this planet, tending to replace them with another artificial world. In the end it severely harms human race by producing exponential inequality and injustice.

How long could we stand with all these linear aberrations? We already have reached the "globalization" stage, so everything is affected by this "plague".

***

Dr. Iain McGilchrist, an ex-neuroimaging researcher at John Hopkins University, and a Fellow of Royal College of Psychiatrist, UK, published a book (2009) titled The Master and His Emissary (subtitled The Divided Brain and the Making of the Western World). Of issue for this text is his idea that *"the mediating/inhibiting influence of the corpus collosum between the two brain hemispheres has become weakened, allowing the logical, linear left to dominate over the sensory panoramic nature of the right."*

As appears, the main role of the corpus collosum is to inhibit the dominance of one hemisphere over another; in my opinion, this role has been changed only in the last 7,000 to 4,500 years; as a result, the left brain became prevalent.

However **this author supports that it should be the other way, where the right brain must be naturally in lead** (I did not read, yet this book, but only the introduction, so I do not want to insist on something I do not know). However, this book is famed as the most important writing on brain lateralization and provides the story how the left brain has created the Western world. He sees a big danger in this process that is unnaturally led by the left brain.

Since this occurrence took place (in the last 500 years), the world of this planet became divided into Western and Eastern cultures, which are antagonistic because they favor the dominance of one single hemisphere (Western has the left, Eastern has the right).

***

In January 2012, the Imperial College London published two studies investigating the effects of psilocybin (a hallucinogenic mushroom) on human brain functioning.

Professor David Nutt, the senior author of both studies said: *"Psychedelics are thought of as mind-expanding drugs, but surprisingly, we found that psilocybin actually caused activity to decrease in areas that have*

171

the densest connections with other areas. We know now that deactivating these regions leads to a state in which the world is experienced as strange".
"The function of these areas, the medial prefrontal cortex (mPFC) and the posterior cingulate cortex (PCC) is proposed to have a role in consciousness and self-identity".

The study found that psilocybin reduces blood flow in certain areas of the brain, while increases the blood flow in others; in fact the psilocybin influences the level of oxygen and glucose level in the blood, and the ability of blood to provide these ingredients to neural networks in the brain. It was found that under psilocybin the activity in more primitive areas become more pronounced (like for the emotional states), but in the main areas the activity drops significantly; even then all aforementioned areas work simultaneously.

In my opinion these studies shown that psilocybin suppresses much of the activity of the left hemisphere, and in the process is removed the ego, or the self-identity. It is known that the left brain produces individualistic approaches, hence the diminishing, or even vanishing of this approach points out to the suppression of the left hemisphere.

The result is comparable with studies on meditation, where the self-identity vanishes, too. Psilocybin and other plants with hallucinogenic effect have been used by humans since Paleolithic. Their magic supported a type of collectivism as a result of the dilution of the self, favored the mystic approaches and interpretations, and also favored more emotional behaviors, which all belong to the right brain. All currently disappearing behaviors and characteristics are those of the right brain. The psilocybin bring them back, but in an altered form because the process mixes information flows from some areas, which currently are dormant, or of very little use.

In other words, the administration of psilocybin changed the chemism of the blood that in response changed the functioning of the brain; in the end, **the proportionality between these functions changed**, and the effect was interpreted as a hallucination (like day-dreaming).

In the course of time, humans used different plants, which influenced the chemism of the brain and its functions. Probably while possibly some of them gave rise to magical functional proportionalities, like some golden ratios.

***

In fact contemporary science allows us to have a glimpse into the idea of the **universal proportionality.**

How much ultrasound energy we need to produce cavitation (that is an implosion)?

What ingredients we need to add to the imploding subject to produce transmutation?

What is the proportion of catalysts (substances) capable to produce LENR?

How much ultrasound can we use to heal bones and wounds?

How much we can amplify or diminish the magnetic field in order to obtain biological and other useful effects?

How much can we dilute a substance in order to produce homeopathic effects?

Could we use homeopathic principle of dilution to manufacture some needed substances in a sustainable manner?

How much we have to stimulate the sinks on the skin (acupuncture) to restore naturally set functioning of the human body?

What is the right proportion of ingredients to produce an artificial stone?

What is the right proportion to create Traditional Chinese Medicine products?

What is the proportion between brain rhythms that allows the best adaptations because adaptation in itself measures what we call intelligence?

What proportionality can stimulate the most profitable brain rhythms?

What proportionality can stimulate nonlinear thinking?

***

The last two questions may already have a partial scientific answer, and it is the simulation of gamma waves, which allegedly create the unity of conscious perception. Gamma waves represent an oscillation between 25 Hz and 200 Hz but with a typical 40 Hz. Gamma waves can be observed in leading a state of neural synchrony from visual cues in conscious and subliminal stimuli. As appears, the neural synchrony produces stochastic resonance in nervous systems.

Andreas K. Engel made the hypothesis that *"the synchronization of neural discharges can serve for the integration of distributed neurons into cell assemblies".*

Protein channelrhodopsin-2 sensitizes cells to light, and during an experiment, the cell activity of these interneurons was manipulated in the range 8-200 Hz. It proved that **gamma waves have a resonating brain circuit property, and that a brain state can be induced by the activation of a specific group of neurons.**

A group study led by Dr. Judson Brewer (November 2011) at Yale University found that *"skillful meditators had decreased activity in the brain's default mode network, which have been linked to attention lapses and disorders such as anxiety, attention deficit hyperactivity disorder, and building of beta amyloid plaque associated with Alzheimer's disease."* The researchers also found that *"when the default mode network was active, brain regions associated with self-monitoring and cognitive control were also activated".* It was suggested that *"the meditators developed a new default mode that's more present-centered and less self-centered."* In other words, meditation *"helped people to stay in the moment that has been part of philosophical and contemplative practice for thousands of years".*

In a 2004 study on meditation among the Tibetan monks, it was observed that **neuronal structures were firing in harmony at a frequency of 25-40 Hz.** The experiment has proved that it **exists a significant relation between visual consciousness and the synchrony arising between many neural areas.** This finding collaborates well with more recent studies, which found in the visual cortex a high gamma activity that is associated with sensory processes.

Numerous studies from '70s were on the scientific investigation of meditation. Most of them came to the conclusion that, **after a long practice, the meditators became spatially oriented and express a great asymmetry difference between visual and verbal skills, leading to right brain greater abilities.**

***

Now we can ask again the same question: were humans in the past able to discover and use some methods to produce harmony between different neural networks of their own brain?

Or, was this harmony the result of externally induced proportionality?

The result may be a yes, or it may be something unknown yet, and unexpected.

25. COMPLEXITY.

For a long time humans observed the behaviors of the environment and of the elements being part of the environment. Common behaviors of many systems or simple elements were considered to be the basic laws of nature. In the beginning this description of the laws was simple, or much generalized. In the process begun to set up goals, which will like to imitate natural behaviors: like melting, freezing, and so on.

I would say humans learnt to domesticate matter, meaning to make matter to adapt to human goals. Hence humans started to artefact matter based on natural while observable behaviors of matter. The development of logics helped to isolate matter from the environment, and to manipulate the material in this isolation until it was obtained the desired behavior.

We have evolved by accumulating knowledge on natural behaviors that became more and more complex laws. Simultaneously we have been able to isolate and domesticate more matter based on progressive accumulation of knowledge on natural behaviors. This domestication of matter is what we call today manufacturing. A big step forward in manufacturing was the discovery of how to syntheses. The behaviors we analyzed are termed physical laws.

All these advents helped us to build an isolated synthetic world, where the complexity became viewed as a mask for simplicity. Is this so?

Almost every element in our environment shows evidence of human artifice. The most significant part of the environment consists mostly of symbols that we perceive as spoken and written language. The laws that govern these string of symbols, the determinants of their content are all consequences of our collective artifice. The very species on which we depend for our food are our artifacts. A plowed field, or a man-made forest are no more part of nature because they became adapted to human purposes. As our aims change, so too do the artifacts. All is analysis and synthesis.

But here everything is governed by natural laws, including the artifacts. And everything is nonlinear in essence. Our artifacted linear systems will only partially respond to vast nonlinearity inhabiting all environments. The linear behaviors of the artifacted are continuously crossed by interferences and interactions of the complexity, where everything is immersed.

***

Any amount of describable details does not lead one to understand a complex system. Depending on the subject, the appearance of an event changes. Thus **we have a diversity of appearances that form and define the complexity**. Complex and non-complex systems interact nonlinearly, and this is common and natural to all systems.

All systems in nature are interdependent. Mathematical systems are closed because mathematics isolates in order to analyze, so mathematical systems are not interdependent because they are independent. Hence **closed systems, or mathematical systems cannot be used to describe nature**.

Coherence is interaction, and the rules of interactions are the fundamental rules of the universe. Coherence defines order, while every order system contains disorder at all levels. The disorder turns into order, and order turns into disorder.

All systems, which are limited, are nonlinear. **All aspects of our perceived reality are nonlinear, being this reality microscopic or macroscopic**.

Linearity is an approximation. The behavior we predict is highly dependent on what we have chosen to assume initially that is our axiom.

***

We are complex systems and naturally nonlinear in essence. Every complex system is formed by numerous macroscopic patterns. Each macroscopic pattern, and its own frequency, are the result of a composition of many microscopic frequencies and entities, but where the macroscopic result is distinct from the summation of its microscopic parts. The combination of many patterns (parts) into a **complex system** makes it to **behave nonlinearly because of its causal plurality**.

Complexity is the theory of Being as Becoming and deals with the **physics of geometry** and topology. Every form in the universe is a scale dependent geometric characteristic. Micro and macro physics depend on the resolution used by the observer when he/she looks at the world. Structures have scaling rules defined by hierarchic properties. The Universe is self-similar from quantum to macro level (fractal geometry). Space-time is a geometric manifold, where every point reveals a structure. Quantum states are complex patterns, being created by the microscopic self-organization processes, which are similar to macroscopic processes. The self-organization process is self-similar and self-consistent at every level.

∗∗∗

According to Illya Prigogine and Nicolis the flow of dynamics of correlation is the deeper truth of complexity. **This "flow of correlations" is the most primitive "non-material" dynamical "matter" of the physical world, producing nonlinearity**, determinism, probabilism and stochasticity **at every level of physical reality**.

At every physical level reality is complex because the development of quantum long range correlations lead to macroscopic phase transition process and **macroscopic ordering. Complexity is a far-from-equilibrium process causing symmetry breaking and macroscopic pattern formation**. In the same manner, the **self-organization** is far-from-equilibrium process that **reveals** entropy decrease and **order development**. Bifurcation that defines symmetry breaking is determined by the geometrical constraints of dimension, shape and symmetries.

Irving Biederman (1987) discovered that every object is composed of primitive shapes or parts, named "*geons*". He found that only 36 "*geons*" are needed to make all objects. However, based on experience an individual can recognize the object just based on 3 "*geons*". This concept is similar in some respects to the Generative Shape Theory.

In language that elemental brick is called "*phoneme*", and for all languages are needed a total of 55 "*phonemes*". The scientists consider that every complexity is the combination and correlation of some *"elemental bricks"*, like *"geons"* and *"phonemes"*.

In the meantime, **images are a collection of *"signs"* that are linked together** in some way by the viewer. The meaning communicated by a *"sign"* depends on another sign. This means that *"signs"* have meaning only within a context. Words and pictures are both collections of symbolic images, where the image is that collection of *"signs"*, and where the combination of *"signs"* form the *"symbol"*.

All these elemental bricks become *"ordered"* into macroscopic patterns by the intermediation of symmetry and symmetry breaking. **Complexity results from the connectedness and the relationship being developed between the elemental bricks.**

Human act of knowing occurs within connectedness developed inside a network. Such **a network is dynamical, relational and spatial and defines the "context".** The connections are naturally and incidentally formed by actions and experience (practice) of the everyday life. To learn is to practice and reflect because knowledge is propositional.

Every complex system is an information processing system (the flows are non-material) and arises spontaneously. **Nonlinear is the rule of complex systems, but it is determined in essence by microscopic processes, where exist a multitude of microscopic causes.**

The superposition of waves with different wave lengths, inhabiting the chaos (microscopic), produces the annihilation of the wave length, and because of that it preserves only the oscillation mode that is called frequency. This is the path from chaos (microscopic) to order (macroscopic). Frequency is governed by resonance that occurs when a system (macroscopic) can store and transfer energy between two or more storage modes, or systems. Resonance is the mechanism by which all sinusoidal waves and vibrations (oscillations) are generated. **The oscillation (frequency) is a type of harmonic motion that is periodic (repeats itself at regular time intervals).**

The fundamental property of the **harmonics** is that they tend to divide the resonating body into integer numbers or equal intervals in a symmetrical, while periodic pattern. Hence **resonance is the outcome of symmetry and it bridges between different frequencies, allowing the information exchange and the energy transfer between patterns.**

Now this outcome defines the **macroscopic that is the direct result of those mechanics, which govern the symmetry**. Because of our own type of brain processing we see the most of the symmetry as a linear aspect. However **linear refers to the macroscopic patterns, which are symmetric**. The microscopic symmetry is governed by some other scalar rules.

Now, going to our analysis, linear complexity is time complexity because an algorithm quantifies the amount of time taken by the algorithm to run the input. On the other hand the algorithm is a set of rules to be followed in computing for solving a problem in a finite number of steps, or **the algorithm is a step-by-step operation, and exclusively linear procedure**. The logarithms are our main mathematical tools in computing all complexities, which are multi-causal and nonlinear in essence. We isolate the content and quantify it. These quantities are subject to mathematics and algorithms; by doing this we eliminate the variables generated by the processes of interaction; we eliminate the natural dynamics.

As appears, the **evolution is only macroscopic**. The attractors of evolution have periodic oscillations, manifesting symmetry and resonance; when the power (frequency) of the attractor changes, this forces the system toward a bifurcation or fork point that allows new orientation toward another but more powerful attractor. In fact here manifests a resonance process (that couples the system to new attractor) and the **new attractor is the manifestation of a different type of symmetry**.

Hence the system itself passes from one symmetry to another. Evolution of matter is mentally recorded as being caused by the symmetry breaking that in fact is the phase transition of the self-organization process. As already explained, it is the transition from one type of symmetry to another.

Preference for symmetry reflects a cognitive appreciation for order and regularity. Our macroscopic world is a temporary island of order (macroscopic) that arises from chaos (microscopic), and it is defined by its various degrees of symmetry, which are intermediated by resonance.

Harmonic resonance is the architectural blueprint of the geometric forms (macroscopic), which are simply the type of various symmetries. This blueprint is flexible, allowing a type of adaptation (geometric adaptation) that compensates for any arising irregularity. When the distortion of the geometry exceeds certain threshold, the pattern automatically breaks into different geometry, encountering that *"emergence"* **that is just transformation.** This process implies self-organization (producing adaptation), and it refers to the adaptive changes in the symmetry pattern. The exterior factors of any system are those factors, which can produce interference (that is an addition) with the system itself, and then the system reacts through adaptation to new ambient conditions by adjusting its symmetry to new patterns.

Complexity results from dissipative (energy entropy) self-organization that seems to gradually downgrade the system's symmetry. As it appears, the **energy itself is define by various degrees of symmetry.** Does the third law of thermodynamics refer to symmetry? Ecological systems are complex dissipative systems (complex symmetries) with mutual metabolic interactions with each other and with the environment. Dissipative systems are irreversible (they cannot be re-built).

The macroscopic phenomena emerge via the nonlinear reactions of microscopic elements of complex systems. The macroscopic stable states originate into microscopic unstable states and modes. Let's say we have individuals at microscopic level, playing the role of parts; at macroscopic level we have the society that is not the sum of its parts (or individuals). The society may be stable but the individuals are not. **The faith of the society is determined by the relations and connections developed between individuals.** If we have a linear society, this is so because of special relationship and particular connections developed long time ago between some members of the society, who eventually have been the core of future elites. **These individuals, when they turned into clusters, were the microscopic elements responsible for the macroscopic linearity within the human society we have today.** The same reasoning applies to previous nonlinear society: the actions of nonlinear thinking members have produced that nonlinear society. In complex systems the behavior of single element is unknown; the same applies to individuals; based on this wisdom we can assume that genetic changes in some individuals with changes in behaviors were responsible for the behavior of the entire group of humans, or their society.

We still wonder what brought linearity to our thinking system? In quantum terms, the collapse of the wave function is a discontinuous change, or the superposition of the quantum system's state (like one inside human brain) with the environment's state (like a severe change in the climate conditions). Or we can say some previously occurred genetic changes were triggered, or activated by changes in the climate at the peak (LGM around 23,000 years ago) and immediate end of the glacial cycle (starting 18,000 years ago). Undoubtable from Glacial Maximum to Deglaciation it was an extreme transition that developed mainly in the temperate zone of the Northern Hemisphere.

As this text has explained until now, we had in the epoch 41,000-12,000 years ago the largest number of geomagnetic excursions, accompanied by the lowest intensity of the geomagnetic field, the highest impact of cosmogenic radiation, which all triggered the LGM and the following Deglaciation. This should be the epoch where we have accumulated a large pool of genetic changes, but which were not active unless a triggering causation occurred (it was a regional element only during LGM and Deglaciation). As we can observe the change in thinking strategy became massive and wide spread only in the epoch 7,000-4,500 years ago, and this epoch precisely corresponds to another geomagnetic excursion named Solovki.

In a way or another Solovki geomagnetic excursion drove us into massive linearity that manifested simultaneously in the most regions of the Northern hemisphere.

There is not uncommon in our current society for a minority to lead the majority, but linearity has been the meaning that provided the original traction for this inequality to develop further on and totally unchallenged. Today we have the extreme case where 1% leads and rules almost the entire world. A mono-polar world, globally, produces the most linear approach everywhere.

But the tendency of "global" development (macroscopic) is the result of nonlinear interactions (microscopic) within a complexity turned global. However **"linear thinking" can provoke global chaos, when it relies only on mono-causality that is an ambiguous selection type of causality**. In this case all other multiple causes are ignored, and their real effects remain uncounted for, or simply distorted by **the paradigm of**

the mono-causality. In this case we will not know which (microscopic) interference might act upon and which attractor will come to dominate our own system. If we do not have **a full pool of adaptations to overcome the result of this nonlinearity,** our available option may not fit in, while such a shortage can produce our own system's collapse.

Complexity contains the wisdom of nature that we ignore at our own peril.

Thousands or tens of thousands of genes determine what we call intelligence. 95% of population scores an IQ of 70-130 points, which is within two standard deviations of the mean. Some 5% of population has 30 points higher intelligence, but some studies have found that up to 70% of IQ score is driven by motivation or even self-motivation. **By contrast lower IQ scores come from lower individual motivation.** This is so because the motivation improves the individual performance, and the motivation is generated by genes (partially inherited, partially newly produced) and the environment.

***

Up to this point, this text has collected a host of information provided by the neuroscience, physics of complexity and other related research. The scope of this text has been to define the main phenomena and particular elements, which have made us "intelligent" and have helped us produce a "civilization" that only occurred during the last 7,000-4,000 years.

Now it is the right time to present in short **the research that come to the conclusion that we are not significantly more "intelligent" that other primates.** Of course these type of research do not make the science highlights, and usually is ignored at the best. The reason I bring these **non-mainstream research** here is because they are the nonlinear alternative to the linear science.

According to aforementioned research, we have the following conclusions (from the *Algernon Argument*):

> -*cortical subdivisions in humans are about as large as expected for an anthropoid primate with 1350 cc brain;*
> -*observations do not encourage the idea that selective pressure for cognition have differentially shaped the proportions of human cortex;*

-humans do not rank first, or even closed to first, in relative brain size (except for the size of cerebral cortex expressed as a percentage of brain mass);

-if encephalization were the main determinant of cognitive abilities, small-brain animals with very large encephalization quotients (like capuchin monkeys), should be more cognitive able than large brained ones (like gorilla); and this is not the case;

-human brain contains some 86 billion neurons (not 100 billion like the literature states) and 85 billion non-neuronal cells (a ratio of 1:1, not 10:1 as reported in literature); human brain has just as many neurons as would be expected of a generic primate brain of its size;

-neuronal densities in the cerebral cortex and cerebellum also fit the expected values in humans as in other primate species;

-the number of neurons in the gray matter of human cerebral cortex, as well as the size of subcortical white matter also conforms to the rules that apply to other primates analyzed;

-human brain has the ratio of cerebellar to cerebral cortical neurons as predicted for other mammals;

-the glucose use by the cerebral cortex or the whole human brain is remarkably constant across the mouse, rat, squirrel, monkey, baboons;

-higher expression of genes related to metabolism in human brains compared to chimpanzee and monkey brains is not directly connected to an increased metabolism per neuronal cell but by lower maintenance of cells in other parts of the body (like cell muscle);

-it is not brain size but, instead, absolute number of neurons that imposes a metabolic constraint on brain scaling in evolution; an estimate for hominid evolution: Australopithecus with 27-35 billion neurons (same number as in orangutan), 62 billion in Homo erectus, 76-90 billion in Homo heidelbergensis, homo Neanderthalensis, and Homo sapiens; the adding of 60 billion neurons required 360 kcal/day;

-the ability to control fire to cook foods may have been a crucial step in allowing near doubling of number of brain neurons in hominids from erectus to sapiens;

-in the last 30,000 years human brain decreased 10%;

-significant aging effects manifest in humans because the leverage of human individuals, who were older that the maximum longevity of chimpanzee;
-once extinct species do not re-evolve; evolution is irreversible;
-**the sum of the number of pathways to intelligence may be quite substantial;**
-**"smartness" is not the necessary fitness of evolution; there is no evidence for some universal selection pressure towards intelligence;**
-primates and humans still evolved towards intelligence because of their social relation and packed structure-there are reasons to doubt that a major evolutionary leap could lead to smarter brains; the evolutionary convergence (that reached maturation) indicates that it would be left little room for improvements; our brain can pack in only so many neurons; neurons can establish only so many connections; these connections can carry only so many electrical impulses;
-evolution tries to compensate for limits in one system by investing more into another system.

All the research mentioned in the previous paragraphs (which come with parallel and nonlinear conclusions) shows that the mainstream research fits well with our **preexisting linear prejudice about the distribution and the meaning of general intelligence in nature. We have our definition on "intelligence" that is "linear" and also evolutionary (that is linear, too), but this forces us into a sequence of obviously consequent linear thinking and linear conclusions.**

***

There is a more or less discrepancy between brain's internal model of the world (that has a nonlinear neural mechanism that must cooperate with the description of macroscopic processes from the environment) and the real environment via self-organization (that is only nonlinear). Linearity of thinking increases this discrepancy to the maximum.

All biological systems deal directly with nonlinear processes and their reaction (microscopic) is a mixture of nonlinear and linear. Nonlinear means the **examination of the starting point**, while looking for the best option to follow. Also the nonlinear generates the **analysis of the context that produces the anticipation of action.**

In humans, the **mirror neurons from Brodmann area 44 enable prediction of what follows** because the mirror neurons are an interface between perception and action. **Mirror neurons can produce "understanding of the context" from information generated by visual system just based on comprehension of the underlying act.**

Structural analysis of music and language is run by the mirror neurons, too. Broca region for musical syntax and language syntax correspond to primary auditory cortex and motor cortex and are processed in parallel. Only as semantics they are processed separately.

***

Western culture differs of Eastern culture by the way the nonlinear complex systems and problems are viewed. **Western culture strives for order, Asian thinking aspires for harmony.** While Western culture ignores or marginalizes nonlinear conclusions, the Asian thinking encompasses and encourages everything that contributes to preserve harmony. Harmony was first developed in mathematics and later in philosophy in the Western culture, while the nonlinear thinkers of the East understood first the nonlinear-complex meaning and applied it directly to philosophy. This differentiation between Western and Eastern cultures arises from **distinct ways to see and understand the natural symmetry**, and the distinction seems to be in the mirror neurons processing.

Cultures are connected with the type of learning that is administrated to the population. Linear approach for learning is very arbitrary because *"the sequences of language items have nothing to do and how learners actually acquire the elements"* and *"the linear sequence is arbitrary"* explains Ken Carroll. He continues:

*"…nonlinear learning is the way that we naturally learned for a couple of hundred thousands of years. In nature linear learning does not exists. We learned instead by doing, through direct experience, through dealing with things as they arose, and through discovering what it was that was important at the time. But most of all **we learned by making connections between stuff we already knew and the stuff we didn't. This meant we actively constructed the knowledge as we needed it. It was very subjective***

and individual *and not linear".* *"… each individual experiences distinct social and psychological phenomena. It is not an objective one size fit, but* *an* *experience that is entirely unique".* *"If anything,* *nonlinear learning* *has more to do with a network than a line.* *It is about experience and connecting the dots…".* *"The web, like the brain, is all about links and nodes. Its* *network qualities have massive implications for learning … and is based on a network of connections …".* *"… the content has nothing to do with your life, your experience, or with anything you might actually want to learn. The effect is one-size-fits-nobody".*

Nonlinearity deals with "context", but not with "content". Context is defined as those *"parts of something, which clarify the meaning"* given to the whole. Or as *"the circumstances that form the setting for an event, and in terms of which it can be fully understood"* (from Oxford Dictionary). Content is defined as *"willing to accept something"*, or *"something that is contained in"* something else, like in a book. (Oxford Dictionary).

***

Complexity can be seen in two ways. The first way is "**contextual complexity**" that provides an *objective* perspective on the realities of the context. The second way is the "**experienced complexity**" that describes the realities from the *subjective* view of the practitioner. *"For practitioners the starting point is the natural context and people's experiences and perceptions of it".* Here the modern scientists try to find an objective "contextual complexity", while the practitioners can find only the subjective view.

This subjective view of the "experienced complexity" was the hominid and human view of those individuals living from Paleolithic to Early Antiquity eras and was summarized by a process of "understanding" of the natural complexity. This pre-historic and ancient "understanding was contextual" (based on understanding the circumstances that form the setting for an event) **and would differ significantly of what contemporary science tries to find out from the "contextual complexity".**

However this "experienced complexity" (and its subdue nonlinearity) has produced all of the old age marvels, which cannot be explained today by our current, mainly linear, mode of thinking.

Based on contrast, perception distinguishes signal from noise and foreground from background. In other words, **perception makes selections from the context but not from the content. In the same way it may "observe" or "ignore" the "content of the ever-changing natural symmetry**", but it will pay attention to that context, where the function gives rise to form (geometry and symmetry).

Good proportion adds harmony, symmetry or balance among the parts. The monotony of symmetry is broken by the addition of a little bit of asymmetry, and here again the role of the mirror neurons is paramount in this type of processing.

***

Symmetry plays an important role in natural selection because within nature the asymmetry is a sign of illness, threat or danger. It is a neuronal basis for consonance (when the curves go in parallel), and here dissonance produces resonance and locked-in frequency.

Symmetry is a macroscopic outcome of microscopic nonlinearity. It is estimated that at the microscopic level it acts the super-symmetry. However **symmetry produces dissipative hierarchies of complexity**. Breaking the symmetry appears as evolution, but in fact it is the phase transition from one type of organization to another, and it occurs during the dissipative self-organization processes. At macroscopic level, **self-organization produces symmetry**. During adaptation, a system runs from one symmetry to another. Here breaking the symmetry precludes the transformation, or the emergence. A **new outcome is new symmetry**.

So the nature of the macroscopic has the tendency to organize the systems in symmetric patterns or configurations. Nonlinear defines many optional points from where the emergence may occur; it is the search for the optimal starting point, or the optimal emergence. **Every complex system is define by its own type of complex symmetry that makes its own coherence**. Coherence is closely related with symmetry, harmony, rhythm, balance, and all their synonyms.

The research of human brain has shown that neuronal **synaptic symmetry increases the network coherence**. Global synaptic coupling can lead to network synchronization with strong coherence. Increasing the number of synapses of the network first leads to global coherence, but when the network becomes very large the system coherence decreases. In the meantime, synaptic asymmetry deteriorates the coherence.

Genus homo belong to the same complex and nonlinear category, and during the millions of years his brain has been trained toward "visual/space thinking". The introduction of language has gradually reduced this ability of "visual/space thinking", tending to replace it with "verbal thinking".

Computing seeks prediction but **nonlinear reality is not computable because it is governed by the ever-changing variables of the context**. We can only compare different systems with different symmetries by the self-similar approach. Cell's software contains programs for **adaptive responses** (by creating pools of genetic mutations) **and they (the pools) are made ready for multiple options because the exact occurring option cannot be predicted**.

However all matter stores quantum information and every elemental particle is a processor of quantum information. The world is a giant quantum computer that proceed with *"quantum computation"* that is essentially distinct of our current means of computing. **We actually sequentially compute the linearity of the macroscopic.**

Manmade quantum computers will provide new type of information processing because the essential features of the quantum world are:

-superposition of quantum states;
-possibility of entangled states.

Eventually the manmade quantum computers will be able to compute the nonlinearity, but currently all computers are still linear (serial computing).

***

Let's see the criticism of our linear world brought by the nonlinear thinkers.

At present organic community relations, which are nonlinear by essence, are converted into **market relationships, which are linear**. Market oriented society has transformed 35% of non-polar lands into human habitats, while men are converted into commodities. Society presses everyone into the blind forcing of conformity that is linear conformity.

All nonlinear behaviors, which are still inherited by many individuals, are constantly suppressed or at best marginalized.

There is no doubt that we currently experience the dictum of linearocracy that is led by the linear thinking that intends to eliminate the nonlinear avatars as anti-social. It manifests into society at large, but mostly in science that is supposed to provide the needed material proof to legitimate the governance and leadership. **Linear science outcomes are taken as the undeniable scientific proofs, which trigger the implementing of social linearity at all levels**.

Every domain of science that, by its research results, may disrupt the market oriented economy and society, is suppressed, ignored, marginalized and/or eliminated if possible. Everything is converted into a homogeneous linearity, where the financial domain, as the sole beneficiary of linearity and market society, reins, leads, diffuses information, manipulates, while it controls the entire social mechanisms.

We can call this the "*linear democracy*", where the "*nonlinear democracy*" becomes the public enemy. Or we can say **we have today the "*verbal democracy*" as opposed to the "*spatial democracy*"**. The roots are in our "verbal thinking" approaches.

Unfortunately this **"verbal thinking", strongly driven by the verbal part of the IT, will eventually and gradually eliminate, or convert more neuronal circuits, turning the brain processing more and more linear**. Even then, the current drive is said to be toward "new visual", and this is so because of the **visual course taken in parallel by the Information Technology. Would the promised Visual Revolution save us from linearity?** As long as the main financial interests are linear, probably this "revolution" will be faked in a way or another. As appears, we have an "ersatz" (replacement) of reality, and we will continue to have it.

Several science papers are currently titled: "*humans are getting dumber*". Even then these papers are an extremely small minority and are mostly unknown by most scientists, marginalized, ignored, and so on. Probably soon they will turn suppressed. Already the mainstream reacts by bringing opposing arguments, and this will continue until the day this idea will be considered non-scientific and will be discarded.

Probably the "*complexity (complex systems) theory*" will gradually become "*linearized*" and incorporated into mainstream "*linearity*". Nonlinear thinkers of the near future will be "the illegals" or the "terrorists" of those future eras. Already "nonlinear thinking" is currently promoted as some manipulation of graphics, and nothing else; it is not connected with networking as it should be; it tends to be inserted in general education as the "*graphic aspect*" (but not networking aspect) of "*linearity of thinking*". Future visual brands are described by new "*visual vocabulary*". Everything is only in connection with the "*content*", while the "*context*" is vastly ignored.

However here there is the hope that a volens-nolens rebound of visual-nonlinear thinking may naturally come to us in the oncoming decades because the computer technology evolves itself in this direction. **It may be a natural cyclicity at work that drove us to linearity and control, while later, as another natural cycle, the nonlinear of complexity will find a way to succeed in our thinking**. I am not sure that this will be the case.

***

In fact our civilization is a startup for dealing with the macroscopic world; but this startup does not help us to fully understand the surrounding reality that is complex. We live with an illusion of control and achievement. But **in many ways we are wrong about everything because the rules of the macroscopic world are made at microscopic level, and we do not understand this microscopic level**. Our species exhausts all resources of the environment, leading to the foreseen environmental collapse, and it also produces an extreme stressing to human wellbeing that nears an extinction possibility.

Our startup originated less than 20,000 years ago that is a very short time period at the level of the natural history. Nevertheless we still are in the first phase (linear phase) that we have briefly started 7,000-4,500 years ago, or even less.

We still are the same primitives but with a language and some illusion of control. It will not be an exaggeration to say that this type of startup may have occurred many times in the last several million years, since a language possibly have arisen among other long extinct branches of different hominids, who all ended in a full collapse. As long as evolution is irreversible, this means, the evolutions of the past have dealt with different types of hominids, or even different primates, who completely perished.

191

As it appears, **the scope of evolution is not to create "intelligence",** and the genetic designs are far from being maximally effective for specialized functions; thus, **cognition-related operations in humans were not significantly enhanced by evolution.**

On a larger cosmic scale we are too primitive to account among the possible long-term civilizations of this universe. When we will be able to understand the complexity and its microscopic mechanism only then we can say that phase two (nonlinear phase) of our civilization started. Only phase two can assure our sustainability and survivability as the whole.

***

Now I like to comment on Albert Einstein article on Religion and Science, published in 1921 and included in the book The World as I See It that was published in 1949 (page 24-28). Somehow A. Einstein gives a philosophical thought on the existence of "polar attributes". In this article he indicated that things are of two kinds: those that exist and those which do not exist.

In fact this is an opinion on polar attributes: those things which exist are the positive ones, and those which do not exist are the negative ones. In sum, he talked about being present (as the positive) and being absent (as the negative).

Based on this approach, one can say that the nonlinear is the "reality", while its absence define the linear. Or, one can say he/she cannot see, perceive or compute the nonlinear. For that person nonlinear does not exist; it exists only the linear. Connecting this idea with human evolution, one can say that the gradual diminishing, or even perishing by suppression, of the original nonlinear thinking, gave room to the development of linear.

In our linear thinking we have expressed this approach by quantification of every analyzed content, and here zero define the border between positive quantities versus the negative quantities. Even quality became quantized in a way or another. Now, the quantities are the building blocks of our world, while they are the product of decoherence, or some illusory products. Having quantities we can build the world step-by-step but from illusory bricks. Our building is intended to produce something similar to the reality, but in fact the real-time result is no match for the real-world because it is made from macroscopic bricks only; in real-world the bricks are made at microscopic level and in a nonlinear/complex way.

Since negative quantities define the "absence" of the positive ones, the "linear" is this absence, or in other words an illusion, or an artificiality, or ersatz (replacement of the genuine one). However this is not a true absence (because it exists multiple layers of reality, and we are not aware of the deeper levels), but for our mind it looks like an absence. Since our brain begun to suppress the nonlinear thinking, in this void was constructed the illusion of linearity. This seems to be the way the human mind works, where we worship an illusion or another.

***

The myths are considered to be the fractal feature of the past events. Here I see an interesting connection with the Hindu tradition that divides human evolution in four stages, or Yugas: Satya Yuga, Treta Yuga, Dwapar Yuga and Kali Yuga. This last stage, or Yuga, started 432,000 years ago that geologically corresponds to Marine Isotope Stage 11 (MIS 11) that occurred during the interglacial placed 424,000-374,000 years ago. It was the longest and warmest interglacial interval of the last 500,000 years.

According to Hindu tradition these four stages symbolize some phases of involution, where humanity gradually lost the awareness of natural reality that was nonlinearity. It is saying that in Treta we lost ¼ of the truth, in Dwapar ½ and in Kali we are left with ¼ of the truth. I do not trust the exactity of this information because as every myth it is self-similar to real events, but as this text has shown, we still have at present some ¼ of nonlinear thinking.

Another coincidence refers to the beginning of Kali Yuga in 3102 BC, and its duration for 5,000 years. It coincides also to Mayan Calendar that defines the origin of the Fifth Great Cycle in 3114 BC, being given a duration of 5,000 years, and the Noah Flood placed 3600 BC that originated into Mesopotamian myths.

In a way or another these ancients were aware of something new starting in the 4th millennium BC. However these dates correspond with a massive implementation of "linear thinking" in those societies. In all three cultures this new beginning was described as a corrupt era that was supposed to end into a Golden Age. Especially the Hindu define this aforementioned new era as an illusion. Many other similarities were expressed in the Chinese philosophy in the form of Taoism (Daoism) and later by the Buddhists.

The Hindu allegories and ideas are the most interesting, and they may suggest some sort of past reality that may be further distorted by modern interferences and interpretations. However they describe a complex past for which we have no proof or clue, yet.

Time and again, the alchemy, transmutation, artificial stone making, the essential meridians and many other things described by the old myths have been proved by the current science as genuine possibilities and approaches of the past. It was there a complexity that we do not understand.

As an example of another complex evolution, we can cite the discovery of 3,172 groups of prehistoric cliff painting at Damaidi, Ningxia (China), which date from 20,000 to 16,000 to 10,000 years ago. The symbols discovered here were ancient Chinese characters, and many the hieroglyphic and abstract signs had acquired the key elements of ancient writings found on pottery, but dated much recent, or 7,000-3,600 years ago. These symbols predated by more than 10,000 years the East China symbols, or the Tartaria tablets from Romania and Balkans. Even another stone from the same Ningxia was carved 30,000 years ago, or 23,000 years before the Tartaria tablets, and 25,000 years before the Sumerian cuneiforms.

There is obvious that **human evolution** in the Ningxia **was quite complex with stages of progress, regress, and discontinuity**. In the meantime it **represented a typical nonlinear approach, where the restarting was the rule of that type of thinking**. Eventually the craftsmen were different people in various stages of transition from Paleolithic to Mesolithic and Neolithic.

What was the connection with the human evolution on the continental shelf of the East China Sea, or with other regions of Central Asia? Of course **the answer in itself is "complex" because it represent a complexity of evolution based on nonlinearity of thinking**.

Somehow similarities of the same nature appeared at Visoko Pyramid Hills (Bosnia) with artifacts dated 30,000-24,000 years ago, and at Gobekli Tepe and Karahan Tepe (Turkey) with artifacts from 11,000-9,000 years ago. The same pattern of building, abandoning and re-starting is present here, while the material connection with the neighborhood is inexistent. These last three cases may be the first approach for the use of artificial stone.

26. Could we have
a rebound of
visual-nonlinear thinking?

Let's first see what the artificiality we have created is?

1. Linearity means proportional changes, where the output is proportional to input (simple cause-effect).
2. Linearity assumes that the sum of the parts is equal with the whole.
3. Linear mathematics brake the behavior of the whole into proportional parts, and these smaller parts are quantified, resolved as numerical equations and added back to the system based on the assumption that they equal the whole (that is a wrong assumption).
4. Analytical mathematical rules are therefore not reliable guides to real world that is nonlinear, and the sum of the parts is not equal with the whole.
5. The linear mathematics only approximate because the system's variables cannot be isolated from each other, or from the context, in order to form a static, or non-dynamical system.

As we have seen in this text, in nature the wave diffraction shows **specific ratios** (see cymatics). These ratios have been named pi-s, or otherwise, and they are no match to quantifiable proportions we have employed in arithmetic and mathematics. As it appears, in nature, **various RATIOS are combined to generate a natural system**, like a plant, a crystal, etc. These ratios are present in all living and non-living systems, and eventually show how these systems (complex and non-complex systems) was created.

One can assume that nature uses a computation type that deals with ratios; **combining different ratios**, where probably exist some ratios used to combine the other ratios, it **is created a complex system**.

Geometry is linear, but it originates from ratios, which are nonlinear. We learnt on geometry from wave diffraction, and from geometry we created the arithmetic and equations. We isolated from complexity simple geometries, which turned quantified.

The ratios create macroscopic products, which behave symmetrical and complex. They contain in themselves levels of order and disorder, which are in a continual dynamic process.

We succeeded to maneuver within the macroscopic environment by introducing linear inputs of our mind strategy, which have produced linear outcomes (outputs). This means the **microscopic and macroscopic environments have been enough stable to allow our constructs to last for sometimes**.

Whatever we have constructed it has been in isolation from the environments, or **we have created some closed systems inside a world of open systems**. All open systems are subject to the dynamics of change produced by interactions. Our closed system outcomes, or constructs, remains in environmental isolation. In other words, **we have created like a boat that floats on an ocean of flows, turbulences and changes**: we are inside this boat that is subject to sea currents, weathering, storms, and can be menaced anytime by some sea life forms (like sharks, whales and some unseen yet dragons).

Our boat is rigid, but it still interacts with the environment in the sense that the boat is subject to a lot of forcing pressure from sun rays, sea salt corrosion, storms, rains, wild sea animals, sea currents and many unseen and unexpected factors, like tsunamis, submarine volcanic eruptions, meteorites, extreme storms (flooding the boat), extreme weather (extreme heat and/or freezing), etc.

In all this we cannot predict what will come, and what the future will be.

The Stone Age people sailed in their boats along rivers, at the sea between islands and coasts, and between continents. They succeeded without any of the technology we have today for boat building and sailing.

By comparing these two approaches, one can say that the final result was the same: both boats (primitive and modern ones) reached their destination, fulfilling the scope of the journey.

The question that arises here is: what was the difference in thinking between these so distinct humans? The primitives were nonlinear thinkers, while their boat was a linear achievement. The moderns are linear thinkers and their achievement is linear. The common element in both cases is the linear achievement, or the boat. The scope is the same (meaning to reach a destination). The only distinction is in the thinking strategy: nonlinear vs. linear. It appears to me that this single difference is huge, and it favored the nonlinear strategy giving vision, scope and means.

***

Let's see how the macroscopic world behave in nonlinear ways.

Here I will take some planetary scale examples.

There are 3.4 million submarine volcanoes, representing 75% of all magma outputs of Earth, from which several hundred thousands are active. During the Ice Age cycles the drop in ocean level relieves pressure on the oceanic volcanic floor, allowing submarine volcanoes to erupt more often. A similar role have the geomagnetic excursions, which diminish the geomagnetic intensity. Over time orbital changes produce alternating gravitational pressure over the tectonic plates, triggering volcanic eruptions. Also Moon cycles generate volcanic pulses every two weeks. Eight out of nine eruptions occur during fortnightly low tides.

Volcanic ashes and gases in the atmosphere produce climate changes: volcanic CO_2 triggers warming, aerosol gases reduce climate warming, the weathering pulls CO_2 out of the atmosphere causing cooling.

Changes in the climate temperature trigger snow fall accumulation and alpine glaciation. Changes in atmospheric currents trigger changes in air moisture, causing desertification. Changes in wind speed can cause desertification, too.

Changes in nebulosity and cloudiness causes changes in atmospheric temperature. Ice accumulation produces dropping of ocean levels. Ice melting rises the ocean level, and produces large flooding.

Changes in sea currents directly changes the coastal climate. Local changes in climate behave by linking too many other regional climate.

Solar cycles can cause warming or cooling, and this depends on cloudiness and snow/ice coverage.

There is undoubtedly a huge scientific proof that the air, water and solid Earth operate as a single system. All life forms are made to adapt to such environmental changes, producing evolution. We in our boat try to protect against all forcing, and therefore, **no evolution occurs in our well isolated and fairly protected boat.**

Here our boat is at the crossing of very many interference paths, and we in the boat do not know what and when some major crossing might occur.

***

Goethe said: "*We do not know what we see, we see what we know*". This means, what we see is determined by the **context** within which the observation is made. In different contexts, the same object may have different meanings. The object is data (content), and data (content) is part of phenomena (systems). Phenomena are complex dynamic systems, showing uncertainty and variance. Data are the evidence of the existing phenomena, **but data is unstable, meaning it changes due to dynamic interactions of systems and between systems.** In opposition, phenomena are more stable. Data cannot characterize the system (the sum of the parts/data is not equal to the whole). Learning to understand is **context** dependent (system dependent) and nonlinear, and it refers to understanding the systems and their dynamics.

Nonlinear is dealing with unknown quantities (variables), which are not fixed quantities, and corresponds to a multitude of evolutionary paths available for each system, or nonlinear deals with **the availability of alternative paths of evolution and the choice between these paths.** Again evolution is macroscopic, only.

Our current approach is based on exclusive quantification, and these quantities are used by mathematics in equations and programs. All equations need quantities, all programs deal with equations. We find quantities through the analysis of the content because all contents are quantifiable. The result is linear mathematics. Our mathematics cannot deal with qualities and changing meanings.

The market itself is seen only as a quantifiable content that allows mathematical analysis. But in fact the market, like everything else, is

199

complex and nonlinear. Because of this we cannot have the slightest idea about what a nonlinear market would be.

***

Let's recall some other characteristics of nonlinearity: it is able to transform an **insignificant difference (sensitivity to initial conditions)** into an appreciable one which has macroscopic consequences; below a **threshold of sensibility**, everything diminishes and disappears, while above the threshold everything increases excessively; **only a discrete (limited) spectrum of paths is available for evolution** (the same discreteness applies to choices); **predictions are unreliable and insufficient (development occurs through accidental choices** arising around bifurcation points, while the bifurcation itself is determined by the self-organization; bifurcation is symmetry changing).

Thus nonlinear thinking must be kept in context, must not require any premises (opposite to logical thinking), must make no assumptions (no premises), must have no kind of pattern, and may use the feelings as part of the nonlinear logics. However, this sounds like the formula for insanity, but it is not because it refers to an ever-changing context where the dynamic interaction (including human interaction) is the rule.

***

There is a misconception among cognitive scientists about the working of the brain like a parallel computer. Recent research have proved that the **brain is not a parallel computer but mostly work like an integrated system**. Harris Georgiou (November, 2014) of National Kapodistrian University (Athens, Greece) shown that a **parallel working appears only for large systems of the human brain but not at neuronal level**.

***

Many prominent contemporary thinkers expressed concerns about the heavy using of the Internet, considering that this has eroded the concentration, memory and capacity of deep thinking. This has been so because of the neuroplasticity of our brain that, when learning new skills, it allows the brain to rewire the existing circuits with new connections arising between the neurons.

The research shown that the undisciplined mind that is in a natural state, would be easy agitated, nervous, wanting, fearful, preoccupied, distracted, hopping from one thought to another. However this finding does not corresponds to processes of nonlinear thinking; this is the type of primitive mind, where the context understanding was not present, yet. Context understanding gave to the bearer a short-term and long-term wisdom that in return produced a dynamic perspective.

By comparison the meditating mind eliminates this hopping from one thought to another and annihilates the distraction and preoccupations. The meditation effects were originally introduced by the use of plants with hallucinogenic effects (especially mushrooms), and by the effects of sound in enclosed cavities, like caves and other stone structures.

However the process of self-domestication contributed to changes in human behaviors, making human groups more social oriented. This changed also the nonlinear characteristics of thinking in the direction of inter-human collaboration and affluent social connections. This was possible because of neuronal plasticity, and changed the brain circuitry. Around 30,000-25,000 years ago begun to manifest, regionally only, an improved type of nonlinear thinking that shown the tendency to build tools capable to materialize, or to give shape to the nonlinear ideas.

The researchers found that the Internet amplifies this tendency of skipping, hopping from one thought to another. This type of amplification changes the circuitry of our brain according with the new design provided on the Internet. A similar change occurred in brain circuitry 500 years ago with the introduction of printed books, and the introduction of reading within the education processes. Several thousands of years before that time the introduction of writing also changed the circuitry, and even before that the linear practice has produced first changes in neuronal circuits. All were possible because of the said plasticity, and all implied suppression of nonlinear paths and circuits.

***

How we can design a **context that trigger human interaction**? Such a context should have **an architecture that maximize the condition for**

possibilities, and minimize the condition for impossibilities. This is to enhance the natural associative/nonlinear process of human mind, or thinking of systems and their interconnectedness. This is **thinking in terms of wholes, which imply to pay attention to wider contexts in which the systems are embedded** because most of the time our perception is contextually driven. In this approach each variable is a dimension in state space: for example profitability is one dimension and growth is another. We have to open the mind to **multi-directional and connected thinking**.

In the meantime, visualization is not intended to exclude verbalization that is its symbolism. As it appears, **visual reasoning tends to impose itself as a new way in Mathematics, as new method to see the unseen**.

During 2012-2014 time period, the marketing turned toward higher **visual content**. This market need was matched by substantial changes in the technology that became oriented toward more visualizations. This situation was feed by new research that shown that 90% of information transmitted to the brain is visual, and **visuals are processed in the brain 60,000 times faster than text**. In the meantime videos on landing pages increase average page conversion by 86%. Visual content is social-media-ready because the **image get more engagement than links or text**.

As it resulted from market research every information to be transmitted was better in the form of an image and/or infographics. Visuals express ideas quickly.

This so called "visual revolution" depends on the way we will be able to solve the aspect "***context versus content***". Right now all efforts are made to introduce (iconographic) images into the content, leading simply to a "*visual vocabulary*". This is a very far cry from the analysis of the context because "**visual thinking is contextual and networked**".

Mental imagery is simply one way of constructing a meaningful relation between subjects/concepts and it emphasis the importance of comprehension in cognition. **Context is the interaction** of different objects/concepts/elements placed into an ever changing (dynamical) medium. Contextual information from images helps for pattern recognition by employing a **mental search machine that uses image input**. This mental search machine or system can express the search intent more clearly, or more relevant to the user, if and when a **visual input** is added instead of the regular verbal input.

The **contextual image mental search system** consists of two subsystems: database system and ranking system. The mental processes require **context capturing** that takes place by capturing a set of visual images (visual context) from the spatial positions of the search medium, and it assembles a database of these images.

The images in the database are disambiguately processed in the brain by using a system that combines the associated contexts and relevance between contexts. In the end the brain uses a **ranking and reranking of the elements with similar contexts**. In short this a real but natural mental search machine, where the **visual input makes tremendous difference**.

However the research involved in current "video revolution" tries to copy as much as possible this mental system by the use of algorithms and other techniques. Unfortunately these techniques are step-by-step approaches, or exclusively linear. Here there are no immediate solutions for introducing the context in new technology. It will require a lot of imagination and knowledge on the side of the designers to create visual contexts and networks similar to "natural reality", or like the fractal self-similarity.

The parts of every system can best be understood in the "context" of relationships with each other and other systems. A system transforms input into output.

***

In recent decades were made new theories on exploiting the context, especially in dealing with business issues. In was developed the "**context analysis**", under the name of SWOT that contains the analysis of strengths, weaknesses, opportunities and threats, but also terms, barriers and resources.

Another new strategy is called **Coopetition, or cooperative competition** that increases the profit by cooperation. It **opposes the strategy that promotes adversaries, where both sides profit from conflict, when attacking each other**. This strategy of adversaries exploits the process of self-organization, **where it is thought that the conflict triggers the process of self-organization**. This last strategy was allegedly used during the first and second world wars along with the false flag strategy.

Situational analysis give shape to context and gives meanings to many things, which may show the need for change.

Quality in a context is the quality of information, because an initiative is informed by all the contextual factors that affect its implementation and sustainability, and it is a form of planning. **Here, all along the history, it was used another strategy of falsifying the information (as deceiving the own public and the adversary in general) with the scope to ruin a given context. This is so, because the acts can be understood when in relation to one another.**

As we can see, the evolution of the visual technology seems to drive us back to the abilities of the right brain and visual thinking. This switching may take several decades, more or less. Probably the **education itself** (like the introduction of networked education) **will change our thinking** due to this visual challenge. And then maybe the **entire structure of our civilization will change**, too.

As it appears, the **networked learning** will replace much of current system of education because people are willing to form networks in order to learn faster, better and with very little spending.

Networked learning is based on three elements concept that is named SSS. **Seeking** is finding things out; **sensing** is how we personalize information by personal reflection and putting ideas into practice; **sharing** includes sharing sources, ideas and experience. One of the most important features of this system is the requirement of **human-to-human interaction** that is having conversations (digitally intermediated or even face-to-face), observing, thinking and using information and knowledge in a win-win process. **This type of network becomes a "context" in itself**, and it can be analyzed and exploit in terms of networking.

***

Our brain experience, especially from childhood, silence away unneeded or unused synapses, since it strengthen the patterns or connections that are repeatedly used, making them structures of the brain. **Brains develop differently because of the dominant type of input they respond to.** During brain development, there are periods when particular neuron groups become amenable to stimulation. If exercise is lacking, some mental abilities may be permanently degraded.

As it appears, nature drove us into the peak of visual nonlinear thinking in the time period 25,000-5,000 years ago, but since the rapid development of languages, it drove us into the inferior level of linear thinking (macroscopic resolution). **Our 5,000 years old civilization have risen out of this primitive linear behavior**. Now we are on a verge of another major transformation, where we will eventually become nonlinear thinkers, again, but this time we will exploit a different level of nonlinear approach.

We can estimate that **gradually the benefits of nonlinear thinking will surface naturally, changing the fundament of our civilization**. The market oriented economy will transform into something more sustainable and more ethical, where inequality can be adjusted. In such an eventuality the quality will prevail over the actual quantifying consumerism, and the natural balance of resources and products will be reinstated in a form or another.

The new approach that embeds anew the nonlinear thinking has the potential to gradually eliminate or severely limit the current linearity and its disastrous outcomes. Maybe we will learn how to let the control go away because the analysis of the context would predict what will naturally follow. Several "phase revolutions" are expected to occur in our doing.

Probably we will start to understand much better the microscopic bricks involved in Physics and the Chemistry, the world microscopic interactions and their meaning, the complexity, the Theory of Chaos, and we will advance in understanding the rules of the quantum world. Our nonlinear computation will gradually become possible. Our understanding about the Universe will change significantly, consequently helping us to understand better our own planet and its mechanics. We will be again closer to nature and the natural ecosystems.

However, **since the entire civilization we have is linear and linearocratic, the changes will be disruptive and revolutionary, but the time period of this accomplishment can be quite long. Nothing will be done unless particular changes will occur in our brain, and these changes cannot be manmade**. We can provide some evolutionary pressure by developing the visual technology, but the aspect of mental transformation will come only naturally and contextually.

We will run again into the mental conflict of the visual versus the verbal, and **we cannot control the future mental outcomes (as development of new neurons and networks, and the intervening of natural suppression)**. It will be unwise to make predictions on any of the future evolutions.

EPILOGUE

It has been a long journey, and some of the chapters written here were not necessary, but they presented the original steps I took in order to be able to draw a conclusion.

In EMBO reports (July 2007) Dr. Wolf Singer, the director of the Max Planck Institute for Brain Research, Frankfurt, Germany put the following question: *"Why did evolution create brains... concerned with analyzing linear processes?"* In the rest of the article, Dr. W. Singer tries to explain and justify this dilemma.

I will quote some of the most important ideas from this article, which have importance for my text.

"This inability is presumably caused by our limited cognitive abilities, which evolved in a world in which there was no advantage to be gained by understanding nonlinear complex multidimensional processes. However, as with quantum mechanics, we can indeed observe processes that contradict our concepts of causality and linearity. The reason why we are so inept at imagining nonlinear interactions might be that, as living beings this ability would have been of little advantage to us. So, there would presumably be no selective pressure for the development of cognitive functions that allow us to comprehend nonlinear dynamic processes."

In my text I had presented scientific information that support the opposite. The nonlinear cognition evolved to a certain point as part of visual thinking, and only later (around 4,000 years ago) a particular genetic manifestation led to the suppression of most of these nonlinear abilities. As it appears, the main cause of suppression had been rapid evolution of language (verbal thinking) that caused competition in the use of different circuits of human brain, while it developed the ability to use visual processing circuits for novel speaking processing. This overlapping began to produce suppression.

Archeological evidence indicates that tendencies toward deliberate cultivation of plants appeared some 22,000 years ago in the Middle East and probably in other places undiscovered yet. However the first villages around the rye cultivation were settled on the Euphrates Valley in North Syria (at Tell Abu Hureyra) by the Natufians some 13,000 years ago. Even earlier, the cultivation of wild rice occurred at the paleo mouth of Yangtze River some 15,000 years ago, and it can also be associated with settlement (but not evidenced, yet) of villages (and eventual cities).

I consider that these tendencies have been correlated with environmental forcing factors, triggering the activation of genes located in the brain's gene-pools. This situation generated a **cognition crises, where new "tools" and strategies were needed**. However the crises was solved by the development of the linear strategies and tools, but this event occurred much later, or around 7,000 years ago. It has been a transition from 7,000 to 4,500 years ago (that has corresponded with Solovki geomagnetic excursion) in which the full implementation of linearity developed in most regions of the world.

But before the full manifestation of this suppression, nonlinear thinking, by the intermediation of the cymatics techniques, discovered the symbols and the geometry (the earliest artifact is dated 20,000 years ago and is from China), the mathematics hidden inside the geometry (the Hindu, Mesopotamians and Greeks), and the transformation of symbols into writing (the cuneiforms of Mesopotamians and the writing of Egyptians). Many other achievements were obtained in the same era that eventually ended with the construction of the megaliths (some 8,000-4,000 years ago), where it was used in a precision in execution that is hard to be reached even today. All achievements of this era indicate nonlinear strategies, which continue to be a mystery for our current linear thinking.

In this text I tried to make an educated guess about the evolution of nonlinear abilities, which still can be described today by giving credit to some of the old myths (because of their fractal nature). Coincidently some of the old descriptions from the myths match current discoveries in science: cavitation, its connection to transmutation, energy production, sonoluminiscence, ultrasound for health, etc. Another idea presented here was about the creation of "artificial stone" and its possible use in construction of megaliths, pyramids, temples, etc.

In the Eastern Culture we have strong evidence about the nonlinear thinking that penetrated deep into this culture: the Hindu (Aryan) philosophy, Buddhism, meditation practices, chakras analysis, Yoga practice, Taoist philosophy, the acupuncture practice and the concept of the meridians of essential energy (after the year 2000 rediscovered in Korea as Primo Vascular System that seems to be more important than the blood and lymphatic systems).

These applications contradict the idea that was no selective pressure for nonlinear thinking and solutions. In fact that pressure existed, but it only created a crises of cognition, and as a result at some point (between 7,000 and 4,500 years ago) a **certain selective pressure favored the linearity of cognition as a solution to a particular functional problem developed by too many while opposing genetic activations.**

However the **natural linear solution was applied by nature as a temporary fixing**. It produced linear tools to be used with nonlinear strategies. But humans took advantage of its rich linear features and developed it further, **amplifying everything into the "artificial civilization" we have now**. Gradually the nonlinear strategies vanished, being suppressed by the linear thinking of the elites.

In my opinion the linearity of cognition came as a stimulus, or an "adagio" to help produce the "tools" necessary to exhibit some nonlinear-"context"-concepts present in the "visual (nonlinear) thinking."

In the beginning, the linear tools were used with nonlinear strategies. Somehow here existed a long term scope: to create nonlinear tools. As appears, humanity failed this long term task, and later linearity replaced context with content. Here we may have some archeological evidence that might unveil some *elusive* artifacts, which were the early nonlinear materializations. The *elusive* stands for their nonlinear origin.

Nonlinear and linear mutually evolve in healthy correlation at the inception of this cognitive composition. Nevertheless the amount of nonlinearity varied from one individual cognition to another. We still describe the nonlinear individuals as the geniuses of those remote eras.

Time and again the cosmogenic influence was "regional", and human society also evolved "regionally." Between such "regions" the time gap separating similar achievements can be measured in hundreds and thousands of years (as in agricultural tendency manifested since 22,000 years ago, incipient agriculture since 15,000-13,000 years ago, and its full scale materialization since 7,000-4,000 years ago).

The earliest city in Europe was Lepenski Vir (Serbia) and was dated 11,500-9,500 BC. It may also be some evidence of an early agriculture that may be dated from the same epoch. I also have to mention the earliest Balkan's script, found as the symbols from the Tartaria-Vinca Tablets (found in Romania) dated 7,500 years ago and the first bronze uses by the Vinca Culture (in Serbia) dated 6,500 years ago.

These are all part of the tools used by the nonlinear thinkers of that epoch. Some of these tools (from the epoch 12,000-8,000 years ago) were the result of earlier geomagnetic influences (excursions) with genetic impact on human thinking.

In my opinion linearity was conceived by nature as a possibility to create tools necessary to deal with universal ratios, or proportionalities. We will see later more about that. However, humans took advantage of this possibility and used it in a different manner that was the most facile way.

In the course of time, human nature favored the spread of linearity in the ways we have today that is to help create "work" as a necessity of new agriculture and new dwelling, exchange of new products as trade, and the creation of an associated market; the work created quantities and the need for counting them as well as the need for keeping records; it appeared the necessity for technology, for better materials (like the use of metals), and for new means of transportation; the result of these "linear tools" was the creation of social classes, money, wealth and power.

So everything evolved differently than the promised technology of UNIVERSAL RATIOS (PROPORTIONALITIES). Randomly, few individuals ran into "golden ratios," but they failed to produce a technology associated with it.

The new world experienced the transition from quality to quantity, and all these have been strong forcing factors, affecting human brain and its way of processing. Even then the human brain remained for long in debt to visual-nonlinear thinking that mostly occurred in the right brain. As shown in this text, we still preserve today a nonlinear thinking that is reflected in the rhythms of the brain before one turns into an adult.

It indicates that social and educational constraints gradually implement linearity in the growing individual, but it fully manifests into adults.

However, not all adults are fully linear, and the statistics show that 24-28% of the individuals are being dominated by the nonlinear thinking, while the linear ones are less than 22% of the whole. Between these two categories exists a median one, where right and left brain are almost equally balanced.

But statistically the linear thinkers dominate the society because they have the right attributes for the right places, and their success in life and the progress into elites is assured by social linear rules. The linear thinkers are in the banking and financial systems; they make money easy, and thus they turn rich fast. The linear thinkers are also the politicians, the lawyers, the regulators, the law enforcers, and the military people. In short, they control the society that is made up to match their linear skills.

With the advent of linearity, natural inequality between individuals became amplified, and injustice and suppression became the norm. Even when most geniuses are nonlinear, other less skillful nonlinear members of the society make the most of the poor, or they inhabit the lower classes.

The business of war was invented (around 4,500 years ago) as the mechanism able to augment local wealth and power of the leaders and the upper classes. At this point the linear became the antithetic of the nonlinear, and the nonlinear became suppressed by the ruling elites.

Is this linearity the right world? Is it sustainable? It cannot be right because it opposes natural evolution and natural selection. It is not sustainable either because it collapses the resources of the natural environments.

Is this "artificial world" dangerous for us? In my opinion **this artificial world can be annihilated any time by nature with another solution** that will have causes we will never guess about. We should know by now that nature produces its own nonlinear computations on scales we do not understand, and **our linearity may be at any time rejected for computational reasons**.

Here there is something else to say: this artificial world has only 7,000 years of existence, even then shards of its incipience have developed several thousand years before linearity became the mainstream of human cognition.

One can speculate that a similar "artificiality" has been created many times, or at least several times, during the history of our planet. Many scientists see the language as the main carrier of linearity, but some of them consider that the language that appeared a long time ago became enough evolved only in the last 20,000 years, or less. Hence language as a prerequisite of linear is also very new.

One can say that we had some 20 glacial and interglacial eras in the last 2 million years and as many opportunities for some very brief episodes of "civilization" because the humanoids exist for more than 5 million years.

However, phenomena in nature are cyclic but irreversible, which means each cycle is different. In other words, the "civilization" episodes created by "linearity" were different from each other.

We do not usually think that the glacial episodes were 10 times longer than the interglacial ones, meaning that current elevation of oceans is an exception from the rule of the last 2 million years. Thus the case of the exposed continental shelves was the norm, and most of previous "civilizations" were developed near the sea, and currently are buried under thick alluvial sediments.

The question is: did there appear a nonlinear "civilization"? In several chapters of this text I tried to figure out for what exactly was used the nonlinear at the beginning of "civilization". As appears the nonlinear thinkers were dealing with proportionality, "context," and with the interactions producing certain results. They learned that the interactions hide something special that was a proportionality called "golden rations"; in fact, it was about specific "proportions" from everything, which we can term as universal constants. There was and is a **"right, or golden or universal proportion" from everything;** this right proportion between ultrasound energy added to air or water (or some other fluid) produces cavitation; and cavitation produces transmutation; we knew for ages about transmutation and the people in this business were called the "alchemists"; Plato was researching for golden (magic) proportions in his study of alchemy, but ended up with the fundament of LOGICS; Isaac Newton was researching the alchemy, but ended up with the principles of MECHANICS.

I recall a blog I read on the Internet a few years ago about nonlinear vs. linear. A physicist said in that blog that the linear was a reductionist approach, and a psychologist said the linear manifested as a bipolar disorder, where the illness (of the linear) was encouraged to become the main and official personality, while the true personality (of the nonlinear) became now the illness. However we have schizophrenia illness, where many of the patients are pure geniuses. So linearity has its own geniality, except that it may destroy us as a species.

Today labs around the world can transmute nickel into copper, copper into gold, oxygen into hydrogen, and so on. The Neolithic people blew a special whistle to produce "cavitation," and eventually few of them obtained copper and gold before the exact proportionality was lost.

Andrea Rossi uses today a secret proportionality to create a catalyst that produces "excess energy" (or thermonuclear reaction at room temperature). He is not alone; many other scientists try today to produce such excess energy at room temperature (not exactly room temperature but temperatures less than 1,500 centigrade).

So, how could a nonlinear "civilization look"? As far as we guess now, the main technology will be based on "golden, universal proportionalities", which will be able to produce everything we have today and much more but in a very simple, sustainable way, without today social distortions, like excessive inequality, injustice, suppression and oppression.

As long as our current "civilization" is linear and "artificial", in our mentality any "civilization" must be similar with ours and hence linear, but in the above paragraph I tried to show the opposite.

Maybe here is the place to make an analogy with Genesis that originated in much older Mesopotamian myths; it was the "flood" that destroyed everything (or the nonlinear world); but Noah saved the world, like in "creating a new linear world" by selecting only the "linear ingredients". And this happened exactly 5,600 years ago. However this new world replaced the allegory of the polytheist beliefs with a monotheist one.

Some genetic speculation shown that the bearers of the R1b haplogroup were responsible for introducing the monotheist belief to the bearers of the R1a haplogroup settled in parts of the Central Asia that turned into the Zoroastrism (in Persia) as opposed to the polytheist (shown in the Hinduism) produced earlier by the Vedic Aryans. As we can see, the polytheism was a nonlinear concept, where changing the context generated new meaning (where one deity can change into another, as the context changes).

In my opinion, a "nonlinear civilization" must succeed to the flow of time, and basically its nonlinearity will be something that at present we have no idea what it will be.

Recently scientists begun talking about the universe in the form of a two-dimensional hologram, where the third dimension is encoded in the viewing surface, tricking our eyes into seeing depth that isn't there (this means the space is flat). The correspondence between the three-dimensional world we see and two-dimensional hologram is nearly perfect.

Another idea is that the laws of nature are based on information that is non- material, hence the universe is not a physical exhibit, and all natural computations are non-material in origin.

In the end I like to say, as most of research, mentioned in this text indicates, that "linearity" creates inequality, and unfair inequality is injustice. Continual injustice is oppression and it is created by a power relationship that perpetuates the inequalities. The gap between the rich and the poor increased constantly during this 5,000 years of linear era, with the peak we experience during current global tendencies.

The truth is, there are natural differences between individuals. In the natural world that is nonlinear these differentiations inhabit every complexity, but they are resolved in a complex manner; however we have no idea how nature deals with inequalities, and we do not know what a natural democracy is. Even then we have our own example of Neolithic communities, where the nonlinear thinking was dominant, and where inequalities were treated amiable, or amicable, and where all members of the community cooperated for a common goal. It was a kind of equality. Since linearity took lead in human behavior, around 5,000 years ago, human society have gradually slide into a world of increasing abuses and injustice. This is an incontestable fact.

However extreme, inequality is a barrier in development because it restricts social mobility, choice, while it increases the weight of circumstances. One can say the politics of the excess linearity turns to be an affront and violation to shared values and principles of human rights. Polar wealth does not serve humanity because it only increases the discrepancies between human rights of different people, and where the pole of the poor has basically no rights at all.

Linearity creates "markets" and market-relationships, which are antithetical to democracy. **Linear power contains features, which are incompatible with democracy and basic concepts of decency**.

We live in a linearocracy, where the unelected "linearity", its money and its power dictate over nonlinearity that still exists in the right brain of all individuals. The world's political systems of power force us to use excessively the left brain to our own detriment and further demise.

The side effects of every linear action represent the unknown we steadily have created for the last 5,000 years, or more. The true results of this unknown accumulation remain unobservable, but the accumulation may potentially be very dangerous, eventually leading to our own extinction.

Should I add anything else?

BIBLIOGRAPHY

Sue Colledge, James Conolly-Elsevier-Quaternary Science Reviews, Volume 101, 1 October 2014, Wild Plants use in European Neolithic subsistence economies...

Nicholas Baumard, Pascal Boyer-Institute of Cognitive and Evolutionary Anthropology, Oxford and University of Pennsylvania, US, Washington University, St. Louis, US-Explaining moral religions

Science Daily-Massacres, torture and mutilation: Extreme violence in Neolithic conflicts, University of Basel, August 18, John A. J. Gowlett-British Academy Centenary Project, SACE, University of Liverpool, UK-The Vital Sense of Proportion: Transformation, Golden Section and 1:2 Preference in Acheulean Bifaces

Alec Julien, We Love Philosophy, March 26, 2014-God Created the Irrational Numbers

Avery A. Morton, Massachusetts Institute of Technology, US-Fibonacci Series and the Periodic Table of Elements

Mihai V. Putz, Laboratory of Computational and Structural Physical Chemistry, Biology, Chemistry Department West University of Timisoara, Romania-Chemistry Central Journal, November 12, 2012-Valence atom with bohmian quantum potential: the golden ration approach

J. Wlodarski, Proz-Westhoven-Federal Republic of Germany-The Possible End of the Periodic Table of Elements and the Golden Ration

Mary Meisner, May 15, 2012-Leonardo Fibonacci discovered the sequence which converges on phi

Md. Akhtaruzzman, Amir A. Shafie, Department of Mechatronics Engineering, International Islamic University of

Malaysia-Geometrical Substantation of Phi, the Golden Ratio and the Baroque of Nature, Architecture, Design and Engineering

Iain McGilchrist, Yale University Press 2009-The Master and His Emissary

Francesca Davenport, Imperial College, London-New study discovers biological bias for magic mushroom' mind expansion

Kate Kelland for Reuters, August 13, 2015-Your Brain On Magic Mushroom Is Actually Similar To Dreaming, Brain Scan Study Shows

Dr. Judson Brewer, Yale University, news release, Nov. 18, 2011- Meditation Can Turn off Regions of the Brain (published on Nov. 21 in the Proceedings of National Academy of Science)

Bertini M, Ferrara M, De Gennaro L, Curcio G, Moroni F, Vecchio F, De Gasperis M, Rossini PM, Babiloni C,-Epub 2006 Oct. 27-Directional information flows between brain hemispheres during presleep wake and early sleep stages

Paleoterran, December 2, 2010 (/journal/2010/12/2/Paleolithic-hunter-gatherer-sleep.html)

J. D. Moyer, Nov. 28, 2014-Overstimulation and Desensitization-How Civilization Affects Your Brain

Adam Chuderski, Krzysztof Andrelczyk, Cognitive Psychology, Volume 76, February 2015-From neural oscillations to reasoning ability: Simulating the effect of the theta-to-gamma cycle length ratio on individual scores in a figural analogy test (on ScienceDirect journal)

Gyorgy Buzsaki, Nikos Logothetis and Wolf Singer, October 30, 2013 in Neuron Perspective, Elsevier Inc.-The Neuroscience Institute, Center for Neural Science, School of Medicine, New York, Max Planck Institute for Biological Cybernetics, Tubingen, Germany, Imagining Science and Biomedical Engineering, University of Manchester, UK, Max Planck Institute for Brain Research-Scaling Brain Size, Keeping Timing: Evolutionary Preservation of Brain Rhythms

Kai Miller, University of Washington, Thilo Womelsdorf, University of Western Ontario-Tim Blanche-Jacob Reimer and Nicholas Hatsopoulos-Markus Siegel-Florian Mormann-Nancy Kopell-Michael Breakspear-The workshop: The consequences of brain rhythms in the organization of neuronal computation

Ewa A. Miendlarzewska, Webke J. Trost, Department of Fundamental Neurosciences (CMU), University of Geneva, Switzerland, in Frontiers in Neuroscience, Jan. 2014-How musical training affects cognitive development: rhythm, reward and other modulating variables.

Susanne Schultz, Emma Nelson, Robin I. M. Dunbar-June 25 2012-Royal Society Publishing, in Philosophical Transactions B- Hominin cognitive evolution: identifying patterns and processes in the fossil and archeological record.

Erin N. Cannon, Kathryn H. Yoo, Rose E. Vanderwert, Pier F. Ferrari, Amanda L. Woodward and Nathan A. Fox, University of Maryland, College Park, Laboratories of Cognitive Neuroscience, Children's Hospital, Boston, Massachusetts, Dipartimento di Neuroscienze, Universita de Parma, Italy, Department of Psychology, University of Chicago- Action Experience, More than Observation, Influences Mu Rhythm Desynchronization (published in PLOS)

Jason Gregory, June 6, 2013-The Artificial Human

Tanya Lewis, LiveScience, April 23, 2013-Brain Asymmetrical Shape Reflects Human Ability, MRI Study Suggests

Dr. Pascale Michelon, Feb. 28, 2008-Brain Plasticity: How learning changes your brain, published in Sharp Brains

Kristina Jennbert-Trepanation from Stone Age to Medieval Period from a Scandinavian Perspective

Migraine Research Foundation-Migraine Fact Sheet

Yuri Moskalenko, Gustav Weinstein, Tamara Kravchenko, Peter Halvorson, Natalia Ryabchikova and Julia Andreeva, 2012, (http://dx.doi.org/105772/5088/)-The Role of Skull Mechanics in Mechanism of Cerebral Circulation, published in INTECH

Michael Green-What Are Linear&Nonlinear Thinkers published in eHow

MIT Technology Review, November 5, 2014-fMRI Data Reveals the Number of Parallel Processes Running in the Brain

Vincent Courtillot, Yves Gallet, Jean-Louis le Mouel, Frederic Fluteau, Agnes Genevey, Paleomagnetisme et Geomagnetisme Institute de Physique du Globe de Paris, Oct. 2006-Are there connections between the Earth's magnetic field and climate? Published ScienceDirect-Elsevier

Dimitra Atri and Adrian L. Melott, 2012, Department of Physics and Astronomy, University of Kansas-Biological implications of high-energy cosmic ray induced muon flux in the extragalactic shock model

James Dyke, Complex Systems Simulation, University of Southampton, June 19, 2015-Earth's six mass extinction has begun, new study confirms; published in The Conversation

Richard Gray, 6 February, 2015, Geophysical Department of Columbia University-Are underwater volcanos causing global warming? Oceanic eruptions may have a greater effect on climate than first thought

Mark McGuinness-Yes, The Internet Is Changing Your Brain

Mindfulnet, 2013-Cognition Improved by Mindfulness Meditation

Wei-Chuan Mo, Zi-jian Zhang, Ying Liu, Perry F. Bartlett, Rong-qiao He, January 23, 2013-Magnetic Shielding Accelerates the Proliferation of Human Neuroblastoma Cell by Promoting G-1-Phase Progression

Journal of Mountain Science, May 2004, Volume 1, Issue 2-Paleomagnetic excursions recorded in the Yanchi Playa in middle Hexi Corridor, NW China, since the last interglacial

Sarah C. P. Williams, October 6, 2014-Genes don't just influence your IQ-they determine how well you do in school

V. A. Bolshakov, Faculty of Geography, Moscow, March 22, 2007-Geomagnetic Excursions: A Reliable Means for Correlation of Geological Deposits?

Deke Xu, Houyuan Lu, Naiqin Wu, Zhenxia Liu, 29 April, 2010, Key Laboratory of Cenozoic Geology and Environment, Institute of Geology and Geophysics, Chinese Academy of Science-30,000-year vegetation and climate change around the East China Sea shelf inferred from high-resolution pollen record, published in Quaternary International, Elsevier

Xin-Shual Qi, Na Yuan, Hans Peter Comes, Shota Sakaguchi, Ying-Xiong Qiu, Zhejiang University, Hanzhou, China, the University of Tokyo, Japan, February 2014-Strong filter effect of the East China Sea land bridge for East Asia temperate plant species, published in BMC, Evolutionary Biology

Roger J. Watt, William A. Philips, December 2000-The function of dynamic grouping in vision, published in Trends in Cognitive Sciences

Donald G. McTavish &Ellen B. Pirro, 1990, Netherlands-Contextual Content Analysis, in Kluwer Academic Publishers

Valery Glazko, Tatyana Glazko, Russian State Agrarian University, 2011-Laws of Anthropogenic (Ecological) Disasters

F. V. Bunkin, V. I. Konov, A. M. Prokhorov, V. V. Savranskii and V. B. Fedorov, 1974-Photoacoustic cavitation in water

G.Tezcanli-Guyer, N. H. Ince, Institute of Environmental Sciences, Istanbul, Turkey, 2004-Individual and combined effect of ultrasound, ozone and UV radiation

Ikuko Kitaba, Masayuki Hyodo, Shigehiro Katoh, David L. Dettmanand Hiroshi Sato, January 7, 2013, Kobe University, University of Arizona at Tucson, US, University of Hyodo, Japan-Midlatitude cooling caused by geomagnetic field minimum during polarity reversal; published by PNAS

Lucy Forster, Peter Forster, Sabine Lutz-Bonengel, Horst Willkomm and Bernd Brinkmann, Oct.7, 2002, University of Munster, University of Freiburg, University of Kiel, University of Oxford-Natural radioactivity and human mitochondrial DNA mutations; published in PNAS

V. A. Dergacev, O. M. Raspopov, B van Geel, G. I. Zaitseva, 2004-The Sterno-Etrussia geomagnetic excursion around 2700 BP and changes of solar activity, cosmic ray intensity, and climate; published in Radiocarbon, Vol 46, Nr. 2, 2004

Dr. Jim Sheedy, the book The Pondering Life, 2011

Dr. Jim Sheedy, Early Right Brain Civilizations, 2011

World Cymatics Congres.com

Gyorgy Buszaki, Brendon Watson, 2012-Brain rhythms and neural syntax: implications for efficient coding of cognitive content and neuropsychiatric disease, Published in Dialogues in clinical neuroscience

Ken Caroll, Dec. 13, 2007-Linear and non-linear learning

Winigrad and Lynn, Dec. 12, 1978-Contextual Imagery

DifferVS.com-Difference Between Content and Context

Lesley Kuhn, 2009-Adventures in Complexity

Stuart Wolpert, July 9, 2008-Scientists learn how what you eat affects your brain-and those of your kids

Viatcheslav Wlassoff, July 30, 2014-Vitamin B12 Deficiency and its Neurological Consequences

V. Dyadigurov, Yuriy Vasilevich Ten, P. L. Meshalkina-Siberian Pine Nut Oil

Miroslava Derenko, Boris Malyarchuk, Tomasz Grzybowski, Urszula Rogalla, Maria Perkova, Irina Dumbueva, Ilia Zakharov, Dec. 21, 2010-Origin and Post-Glacial Dispersal of Mitochondrial DNA Haplogroup C and in Northern Asia, published by PLOS/one

Miroslava Derenko, Boris Malyarchuk, Galina Denisova, Maria Perkova, Tomasz Grzybowski, Elza Khuznutdinova, Irina Dambueva, Ilia Zakharov, Feb. 21, 2012- Complete Mitochondrial DNA Analysis of Eastern Eurasian Haplogroups Rarely Found in Populations of Northern Asia and Eastern Europe, published in PLOS

Anatoly Karlin-Analysis of China's PISA 2009 Results

Nature, Nov. 20 2013- Surprising aDNA results from Paleolithic Siberia (including Y-haplogroup R), published in Eurogenes Blog

Nature, Feb. 2015-Eastern Europe a bifurcation hotspot for Y-hg R1, published in Eurogenes

Emily Saarman, May 31, 2006-Feeling the beat: Symposium explores the therapeutic effects of rhythmic music, published in Stanford News

Ana Trafton, MIT News Office, Sept. 27, 2011- Brain Rhythms Are Key to Learning

Michael Wei Liang Chee, Hui Zheng, Joshua Oon Soo Goh, Denise Park and Bradley P. Sutton, Duke-NUS Graduate Medical School Singapore, University of Illinois at Urbana, University of Texas at Dallas, April 23, 2010, published by J. Cogn Neuroscience- Brain Structure in Young and Old East Asians and Westerners: Comparison of Structural Volume and Cortical Thickness

Li M, Luo X J, Reitschel M, Lewis C. M. at all, Bipolar Consortium, Swedish Bipolar Study Group-Allelic differences between Europeans and Chinese for CREB 1SNPs and their implications in gene expression regulation, hippocampal structure and function, and bipolar disorder susceptibility, published by Nature Publishing Group

Henrik I. Elsner and Erik B. Lindblad, Dec. 1989-Ultrasonic degradation of DNA

Walter J. Freeman and Giuseppe Vitiello, University of California, Berkeley, Univ. di Salerno, Italy, Nov. 2005-Nonlinear brain dynamics as macroscopic manifestation of underlying many-body field dynamics

Martinovic J., Busch NA, 2010, Elsevier-High frequency oscillations as a correlate of visual perception

Jared Diamond, University of California at Los Angeles Medical School, May 1987-The Worst Mistake in the History of the Human Race

Jared Diamond-Evolution, consequences and future of plant and animal domestication

Clive Gamble, William Davis, Paul Pettitt, Lee Hazelwood and Martin Richards-The Archeological and Genetic Foundations of the European Populations during the Late Glacial: Implications for Agricultural Thinking

Yi Rao and Jane Y. Wu, Washington University, School of Medicine, St. Louis, Missouri-Neuronal migration and the evolution of human brain, published in Nat Neuroscience, Sep, 2001

Cell Press, Feb. 21, 2015-Newborn neurons in adult brain may help us adapt to environment

Jane M. Healy, PhD- Endangered Minds

Geek Buffet-Visual vs, Verbal Ways of Thinking

Temple Grandin, Dept. of Animal Science, Colorado University-Thinking the Way Animals Do: Unique insights from a person with a singular understanding, published in Western Horseman, Nov. 1997

Temple Grandin-Animals in Translation

Charles Brack-Conservative Left Brain, Liberal Right Brain

Stanford Encyclopedia of Philosophy-Dual Coding and Common Coding Theories of Memory

John Harris-The Brain of the Denisovan Girl

Michael Leyton, Springer-Verlag, New York, 2001-A Generative Theory of Shape, book review by Stephen R. Wassell

Michael Leyton-Symmetry, Causality, Mind, Bradford Books, 1999

Dr. Gil Stein, Oriental Institute of the University of Chicago, April 2010-What Happens When Mobile Hunter-Gathers Settle Down

Evan Ratliff-Taming the Wild

Daily News-Sound Phenomena Influenced Ancient Art and Architecture, Say Researchers, published in Popular Archeology, winter 2015

Daily Mail-14 May, 2012-A New Chapter in Human History: Starling discovery of Stone Age caveman in China who are an entirely new species

Ken Kurzweil, Viking Press-The singularity is near

Time, in Science, Nov. 15, 2012-Human Beings Are Getting Dumber, Says Study

Marta Lahr, Cambridge University's Leverhulme Center for Human Evolutionary Studies, Feb 3, 2015-Scientists Hit Record by Decreasing the Size of Human Brain

Katy Waldman-Lascaux's Picassos-What prehistoric art tells us about the evolution of human brain

Andrew Moseman, Nov 9, 2010-Studying Neanderthal Brain Development, One (Indirect) CT Scan at a Time

Ed Yong, Nov 11, 2009-Revising FOXP2 and the origins of language

Shangbin Xiao, Anchun Li, Fuqing Jiang, Tiegang Li, Shiming Wan and Pen Huang, Institute of Oceanology, Chinese Academy of science, spring 2004-The History of Yangtze River Entering Sea since the Last Glacial Maximum

Houyuan Lu, Zhenxia Liu, Serge Berne, Yoshiki Saito, Baozhu Liu, and Luo Wang, published in Boreas, Oslo, Vol. 31-Rice domestication and climatic change: phytolith evidence from East China Liangyong Zhou, Jian Liu, Yoshiki Saito, J. Paul Liu, Guangxue Li, Qingsong Liu, Maosheng Gao, Jiandong Qiu-Laboratory of Hydrocarbon Resources, Qingdao-Fluvial system development and subsequent marine transgression in Yellow River delta during last glacial maximum, published in Elsevier, Nov. 2013

Aylwyn Scally and Richard Durbin-Revising the human mutation rate: implications for understanding human evolution, published in Perspectives

F. Javier Pavon Carrasco, M. Luisa Osete, Miquel Torta and Angelo de Santis, Instituto Nazionale de Geofisica e Vulcanologia, Roma,

Italy-Archeomagnetic jerks at global scale during the Holocene period

Dan Eden-Is DNA the next Internet?

Michael Balter, June 21, 2010-Romanian Cave May Boast Central Europe's Oldest Cave Art

Mike Marshall, NewScientist, April 5, 2014-Mystery Relations

Umesh Kapil, Oman Medical College, 2007-Health Consequences of Iodine Deficiency

Science Daily, Oxford University press, Sep 24, 2013-Genetic study pushes back timeline for first significant human population expansion

Science Daily, University of Bristol, May 28, 2012-Inequality dates back (7,000 years ago) to Stone Age; Earliest evidence of differential access to land (hereditary)

K. Kochman and M. Czauderna, The Kielanowski Institute of Animal Physiology and Nutrition, Polish Academy of Science, November 2010-The necessity of adequate nutrition with diets containing omega-3 and omega-6 fatty acids for proper brain development, function and delayed aging: Review

The Columbia Encyclopedia, 2014-Neoteny

Sergio Prostak, Dec 1, 2012-Engraved Stone Dating Back 30,000 Years Found in China

Burchard Brentjes, Feb 21, 1999-Rock Art in Russian Far East and in Siberia

Daily Mail, Dec 31, 2010-Are we becoming more stupid? Human brain has been shrinking for the last 20,000

Ed Yong, June 21, 2011-Humans have a magnetic sensor in our eyes, but can we detect magnetic fields?

John Hawks, prof. of Anthropology, University of Wisconsin, Madison-How Has the Human Brain Evolved? Published in Scientific American

AG. Guskova, O.M. Raspopov, A.I. Piskarev and V. A. Dergacev, Physico-Technical Institute of RAS, St. Petersburg, Russia, 30 May, 2008-Magnetism and paleomagnetism of the Russian arctic marine sediments

Morner NA, Gooskova E, Gernik V, Piskarev A, and Raspopov O., EGS XXVII General Assembly, Nice 21-26 April 2002-Magnetic Properties of Sedimentary Rocks of the Barents Sea

Prof. Dr. Wolf Singer, 2007, Director Emeritus of Max Planck Institute, Germany-Understanding the brain; published in EMBO Rep, 2007, Jul 8

Science Daily, May 13, 2011-Last Neanderthals near the Arctic Circle?

Science Daily, March 12, 2012-Was human evolution caused by climate change

Science Daily, Sep 22, 2010-Neanderthals more advanced than previously thought: They innovated, adapted like modern humans, research shows

Science Daily, Aug 14, 2014-neanderthals overlapped with modern humans for up to 5,400 years

Science Daily, Nov 6, 2014-Ancient DNA shows earliest European genomes weathered the Ice Age: Neanderthal interbreeding clues with mystery human lineage

Science Daily, Sep 17, 2014-New branch added to European family tree: Europeans descended from at least 3, not 2, groups of ancient humans

Science Daily, June 10, 2014-Seafarers brought Neolithic culture to Europe, gene study indicates

Science daily, Oct 18, 2012-Prehistoric human population prospered before the agricultural boom, research suggests

ABOUT THE AUTHOR

Reality is not as is, but as interpreted. Humans are caged, not freed, by the logic of own thought that conceals everyone behind own egocentric self. Emotions and intuition are created by magic, golden irrational that is ratio and balance. The irrational separates number from magnitude, while this magnitude cannot be found. The logical, rational thought, as the sole expression of our civilization, starts from somewhere, follows a logical path, and ends with a conclusion because all ideas have an end in themselves. How close to this conclusion, or end are we? Is the rational world only a distorted projection of the irrational world? Does civilization make sense only for a linear mind?

Printed in the United States
By Bookmasters